Naturalists' Handbooks 20

Microscopic life in *Sphagnum*

MARJORIE HINGLEY

With illustrations and plates by Peter Hayward and a plate by Diana Herrett

Pelagic Publishing
www.pelagicpublishing.com

Published by Pelagic Publishing
www.pelagicpublishing.com
PO Box 874, Exeter, EX3 9BR

Microscopic life in *Sphagnum*
Naturalists' Handbooks 20

Marjorie Hingley

Series editors
S. A. Corbet and R. H. L. Disney

ISBN 978-1-78427-274-6 (Hbk)
ISBN 978-1-78427-273-9 (Pbk)
ISBN 978-1-78427-275-3 (PDF)

Digital reprint edition of

ISBN 0-85546-292-2 (Hbk, 1993)
ISBN 0-85546-291-4 (Pbk, 1993)

British Library Cataloguing in Publication Data
A catalogue record for this book is available from the British Library.

Contents

Editors' preface

The bogland habitat, which is often threatened by peat extraction, has very great natural history value. It is known to support interesting plants, dragonflies and birds. Less widely recognised, but no less important, is the unique community of microscopic plants and animals inhabiting the leaves and crevices of *Sphagnum*, the moss that dominates the bog vegetation. This community is well worth exploring. Under the microscope, a drop of water squeezed from bog moss will reveal a wonderful diversity of complex and distinctive organisms. Some are conspicuous and easy to name; others are poorly known and identifiable only to the major group. This book introduces the community, and gives keys for identification, which has hitherto required the use of a range of rather inaccessible literature and specialist expertise. Perhaps this book will help to encourage research that will reveal more about the range of species present, their natural history, and particularly the interspecific interactions that make the bog moss ecosystem so fascinating to ecologists.

Ecological processes are usually studied in the familiar context of a wood or a meadow. Here they operate on a more convenient, compact and accessible scale. If moss samples are collected with care and restraint from unthreatened sites, an ecological study performed on a handful of *Sphagnum* need not harm the environment and will bring great benefits in terms of increased understanding and appreciation of a microcosm that deserves a wider audience.

S.A.C.
R.H.L.D.
December 1991

Acknowledgements

Dr Sarah Corbet's paper on the testate rhizopods inspired my search for microscopic life in *Sphagnum*. Both she and Dr Henry Disney, as editors, have shown great patience in helping me to organise the results of this fascinating process of discovery. My grateful thanks also to Dr Peter Hayward who has dealt so amiably and promptly with a bewildering variety of material to produce the fine illustrations. The illustrations in Plate 3 are by Diana Herrett (née Harding). I thank her, and the Field Studies Council, for permission to reproduce them here. Mr Alan Eddy has answered numerous questions relating to *Sphagnum* moss as well as identifying species. Drs Colin Curds and Bland Finlay have assisted me in sorting the ciliates; Professor Alan Brook has guided me amongst the desmids; and Mr John Carter, Dr Elizabeth Haworth and Dr David Mann have helped me with diatoms. Dr Malcolm Luxton has shed light on British oribatids and Dr Charles Hussey helped me to compile the rotifer list. I am also grateful for the use of facilities in the Botany and Zoology Department libraries at Cambridge University and for the use of library facilities and the Fritsch collection of Algae at the Institute of Freshwater Ecology, Windermere.

Every effort has been made to use current nomenclature. In this taxonomic jungle errors and omissions may remain even after meticulous checking. For these I apologise and take full responsibility. I wish all my readers happy hunting and much enjoyment in the process!

M.H.

1 Introduction – the bogland habitat

The mosses belonging to the distinctive genus *Sphagnum* are generally known as bog mosses. As the name suggests, they are characteristic of bogs, a particular class of wetland habitat. Wetland habitats are classified according to the presence or absence of peat. Peat-free wetlands with vegetation growing in mineral soil are called marshes. They are rarely colonised by *Sphagnum*. Wetlands with vegetation rooted in wet peat are called mires.

peat: an accumulation of unconsolidated, partially decomposed plant material in a wetland

The nature of a mire depends partly on the quality of its water supply. Mires fed largely by rainwater, which is poor in mineral nutrients (oligotrophic), tend to be distinctly acid (pH less than 5.7). These rainwater mires are called bogs and they are prime sites for bog mosses, with *Sphagnum* species tolerant of high acidity. Mires fed by groundwater which carries nutrients derived from the surrounding land are usually less acid than bogs. Mesotrophic mires (pH 5.7–6.5) may have an abundance of other species of *Sphagnum* that do not grow in very acid waters. Eutrophic mires support few species of *Sphagnum*. Mires that are neutral or alkaline are called fens; *Sphagnum* is usually sparse or absent.

oligotrophic waters: waters low in mineral nutrients

eutrophic waters: waters rich in mineral nutrients

Rainwater mires (true bogs, sometimes called ombrogenous bogs) predominate in most upland parts of Britain, except where slopes are steep. Raised bogs occur on more or less flat areas found in low-lying country, on alluvial plains in broad river valleys or on marine sediment raised above sea level. There may be an encroachment of farmland caused by the reclamation of such places by drainage or removal of peat and the addition of fertilisers. The cut faces of the peat may leave depressions in which secondary colonisation forms wet *Sphagnum* bog again. Blanket bogs depend on high precipitation and are characteristic of the Pennines, and of uplands in Ireland and the north and west of Scotland. The gentler slopes and level ground are covered in an unbroken mantle of peat. Raised bogs and blanket bogs are not distinct categories: they are linked by a series of intermediates.

ombrogenous mires: mires whose waterlogging is due to frequent rain

Bog vegetation consists of a more or less continuous cover of *Sphagnum* species, a special community of small plants which live on this carpet and a somewhat stunted vegetation of flowering plants, mainly dwarf shrubs and monocotyledons, that grow through it. The dominant bog mosses form a continuous carpet thereby raising the whole surface, which grows upwards. As the tips of the *Sphagnum* plants grow upwards, the lower parts die and become peat. Wherever such *Sphagnum* cover is well developed, active peat formation is taking place and the bog is growing. The sphagna form the major component of peat; vascular plants are more irregular and localised.

When sphagna dominate the vegetation they occupy a unique position, forming a living surface to the peat, and are themselves part of the substratum for the vascular plants as well as a habitat for a host of microscopic plants and animals. Typically, bogland vegetation where sphagna are actively growing develops a system of hummocks and hollows with carpets or lawns between. The average difference in elevation between hummock and hollow is about 30 cm, but in some bogs the tallest hummock may reach 70–80 cm above the mean level of the bog surface and the hollows may attain a depth of 100 cm or even lower below it. The deeper hollows lie below the water table and appear as distinct water-filled pools. Areas where the bog surface lacks such micro-relief are called flats, lawns or carpets (fig. 1).

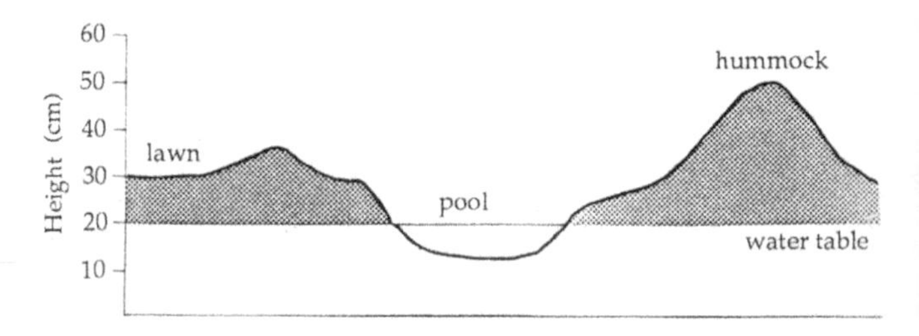

Fig. 1. Diagrammatic section through hummock and hollow system

Hollows are usually elongate in outline and lie with their long axes parallel to the general surface contours of the bog, at right angles to the direction of slope.

The upward growth of the hummock and hollow complex is an irregular process. On some bogs hollows have been converted into flats and hummocks but on others they are persistent features which have occupied the same places for a long time. The process of bog growth is still very imperfectly understood.

Sphagnum species are not limited to the places described above. They may also grow on damp heaths, in acid woodlands and beside streams where the pH is below about 6.5. Bog mosses are best developed in the north and west of the British Isles by reason of climate, topography and less human interference, but there are still good *Sphagnum* habitats in Kent, Sussex, Surrey and Hampshire, many of them on nature reserves.

Sphagnum-rich areas are shrinking fast owing to the demands of agriculture, forestry and peat extraction. The conservation value of many of their plants and larger animals is well known, but the varied and fascinating microscopic life supported by *Sphagnum* plants has hardly been explored. Fortunately, this can be appreciated even in a small handful of moss from a non-threatened area.

2 The *Sphagnum* plant and its physical and chemical environment

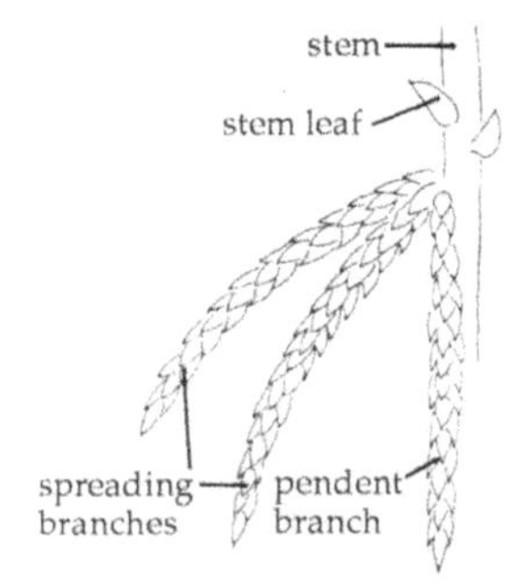

Fig. 2. Spreading branches, pendent branches and stem leaves in *Sphagnum subnitens*

The bog mosses are members of the Sphagnopsida, one of the four main divisions of the Bryophytes. This division contains the single family Sphagnaceae within which the only genus, *Sphagnum*, includes about 150 species. These fall into 11 well-defined groups. Six of these groups are represented in Britain with about 30 species, some of which are difficult for a non-specialist to identify.

Branches arise from the stem in bundles of 1–8. These are differentiated in most species into spreading branches which bear the main photosynthetic leaves, and pendent branches which are pressed close to the stem and may assist in water conduction (fig. 2). The outer cells of the branches may or may not have pores. Those with pores are called retort cells and may protrude from the surface.

Leaves have a very characteristic and specialised structure by which *Sphagnum* is easily distinguished microscopically from other mosses (pl. 2.1). The main part of the leaf consists of a network of two types of cell: narrow living, green cells and inflated dead, clear or hyaline cells which give mechanical support to the green cells. The walls of the hyaline cells usually contain fibrils arranged in a hooped or spiral pattern. The walls may also be perforated by pores which vary in size and shape according to the species and to their position on the leaf. When these dead cells contain air in dry conditions, the leaves become colourless and appear white and brittle; when the cells are full of water the leaves appear green or tinted with colour. They enable *Sphagnum* to act as a sponge, holding up to 20 times its dry weight of water.

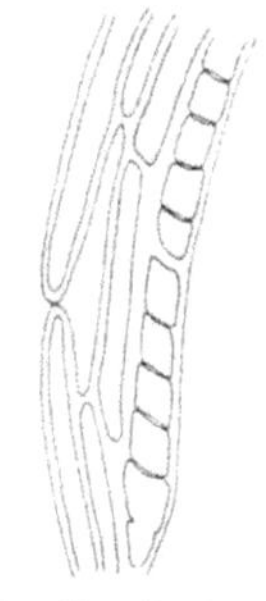

Fig. 3. Leaf border in *S. subnitens*

Sometimes the leaf has a border 1–9 cells wide of elongated cells which die when the leaf is mature (fig. 3).

When leaves are mounted in a drop of water to examine their microscopic structure, a wide range of microscopic organisms may be seen around the leaves, on the leaves or even inside them. A drop of water squeezed from wet moss may reveal a micro-menagerie of such organisms. It is with these organisms and their relationship to the *Sphagnum* plants that this handbook is concerned.

micrometre (μm): one thousandth of a millimetre

Corbet (1973) describes the assemblage of plants and animals living in or amongst *Sphagnum* leaves as unlike the flora and fauna of any other habitat. Nearly all the organisms are small enough to inhabit the film of water in the concavity of a leaf, which may be only 300 micrometres in diameter (fig. 4). They can tolerate the very unusual physical and chemical conditions. These show great variation from the top of the plant to the bottom, and may fluctuate sharply through time. The distribution of micro-organisms often shows a close correlation with them. Temperature, light and humidity are among the most variable factors, but the concentration of oxygen and the acidity (pH) must also be taken into account.

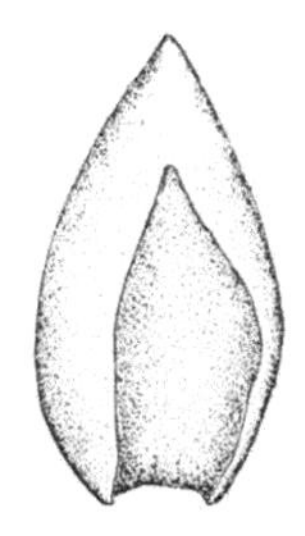

Fig. 4. Concave leaf of *S. palustre*

Little experimental work has been done in this field and there are numerous opportunities for research.

The surface temperature of a carpet of *Sphagnum* fluctuates considerably during a period of 24 hours. The range of temperature depends on such factors as season, rainfall, cloud cover, wind, mist and the degree of shading by other vegetation. The moss is a bad conductor of heat so the temperature changes much more slowly in the lower layers than at the surface. By the middle of a fine summer day a steep temperature gradient is found in the top 10 cm. Towards evening the surface cools faster than the deeper layers and the temperature gradient is reversed. Fig. 5 shows the results obtained by Corbet on the surface of *Sphagnum* at Tarn Moss, Malham Tarn on a sunny day in July 1972. Norgaard (1956) records similar results from a Danish bog and relates it to the habits of two spiders. It would be interesting to see how the vertical temperature profile in *Sphagnum* growing in shaded woodland conditions differs from that in areas exposed to the sun.

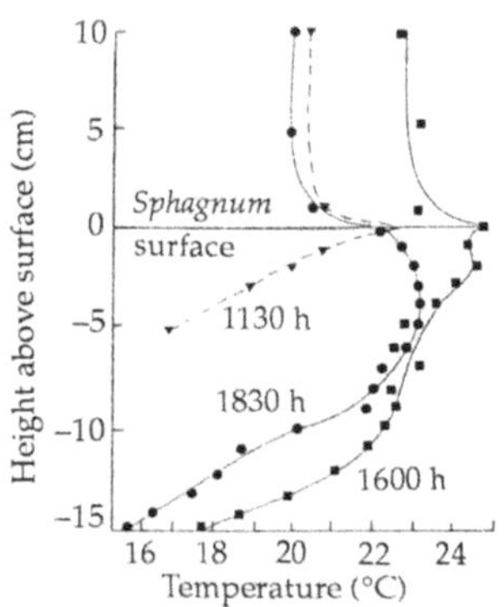

Fig. 5. Temperature profiles across surface of *Sphagnum* on Tarn Moss, at 1130 h, 1600 h and 1830 h BST, on a sunny day in July 1972 (after Corbet, 1973).

Temperatures in *Sphagnum* hummocks may vary horizontally as well as vertically. The daily regime of temperature is probably very different on the south side of a *Sphagnum* hummock from that on a north slope. Differences may also be found between hummocks, lawns and pools.

Light intensity falls off steeply with depth as the *Sphagnum* 'heads', packed close together, intercept most of the light. The region light enough for photosynthesis is therefore only a few centimetres deep (fig. 6). Some aspects of the vertical distribution of testate rhizopods may be related to this (p. 17) and light intensity is expected to be a major factor for other elements of the microflora and microfauna, such as desmids, diatoms, or flagellates like *Euglena*.

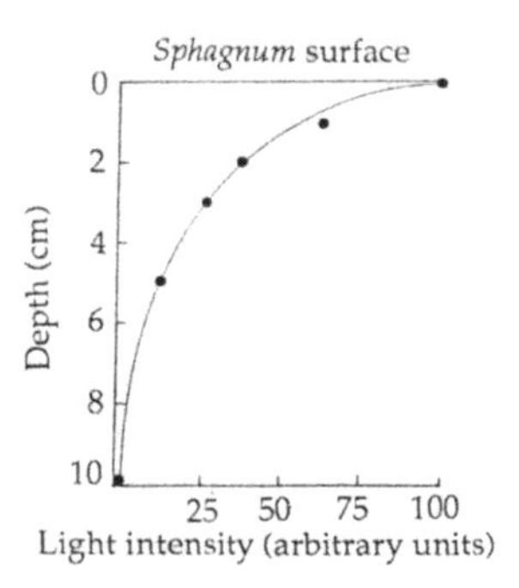

Fig. 6. Change of light intensity with depth in *Sphagnum* carpet on Tarn Moss (after Corbet, 1973).

Differences of humidity are of great importance. A bog pool and hummock may be only a few centimetres apart but offer very different microhabitats. Hayward & Clymo (1982) have shown that the structure of different species of *Sphagnum* plants influences their water content. The larger spaces, which are the main pathways of water transport, are outside the plant cell walls: between leaves and between pendent branches and stems. The mean radius of such spaces around *Sphagnum capillifolium* is smaller than that around *S. papillosum*. For a given depth of water table, the water content of the topmost tuft of branches, where growth occurs, is greater in the former species than in the latter. Hayward & Clymo relate this difference to the position of *S. capillifolium* in hummocks and *S. papillosum* on the bog surface (fig. 7).

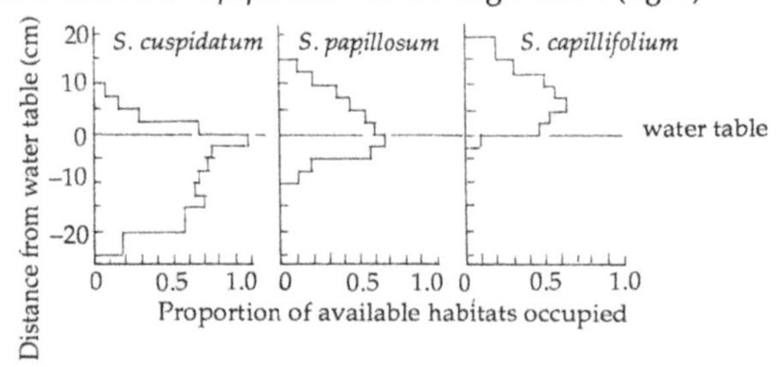

Fig. 7. Proportion of 1 m^2 quadrats containing three species of *Sphagnum* at various heights in relation to the water table on the Silver Flowe, Kirkcudbright (after Hayward & Clymo, 1982).

The effect of this difference on the microscopic inhabitants of the two species would be well worth exploring.

The concentration of oxygen must also show a steep vertical gradient at the surface. The water will be in equilibrium with air, and perhaps locally supersaturated with oxygen, when the moss plants are photosynthesising. In the compact mass of dead *Sphagnum*, about 20 cm down, there must be very little free oxygen, as there is no photosynthesis here and oxygen can only reach the region by diffusion from above. Below the water level, bacteria and fungi use up the oxygen before it can penetrate far. The shortage of oxygen may limit the depth at which aerobic species can live. This in turn limits the rate of decomposition. This is why peat accumulates in waterlogged conditions, but is lost by decomposition when peat fens are drained and ploughed for farming. Field experiments are needed to measure oxygen concentration in bog pools, hummocks and lawns, just below the surface and at depths of, say, 10, 15, 20, 25 and 30 cm and to see how this relates to the abundance of certain indicator groups of species, such as fungi, bacteria, and aerobic and anaerobic protozoa (Finlay & Fenchel, 1989).

If a moss or lichen is placed in a solution of a metal salt, metal ions are taken up and hydrogen ions are released. Ion exchange takes place in a matter of minutes and does not require the expenditure of metabolic energy (Richardson, 1981). *Sphagnum* is remarkable in having large numbers of ion exchange sites in the form of uronic acids produced during the synthesis of new wall material at the growing tip. This results in the release of hydrogen ions and ensures the low pH which is ideal for development of the moss and which helps inhibit competing species (Clymo, 1963).

Sphagnum bogs are acid, commonly with a pH between 3.8 and 4.6. There is a tendency for each *Sphagnum* species to develop a specific pH in the water contained in the hyaline cells. In south-eastern England, Rose (1953) found that the pH in the moss carpets and hummocks varied seasonally. In winter, when there was much rain and little evaporation, metallic ions were leached out from the hyaline cells. This raised the pH (decreased acidity) in the hummocks, whereas in the pools the pH remained fairly constant (table 1).

It would be interesting to see whether this is also true in blanket bogs in the north or in woodland areas where different *Sphagnum* species are found.

Since *Sphagnum* species differ from one another in habitat, pH and water-holding capacity, they might be expected to harbour different assemblages of microscopic animals and plants. Species-specific differences in the nature of this microscopic community have hardly been explored.

Most of the work done so far on the distribution of organisms in relation to physical and chemical conditions has been concerned with the testate rhizopods (Corbet, 1973). These animals are some of the most characteristic of the habitat and are described on pp. 15–17.

The organisms associated with bog mosses occur in great variety and the numbers of individuals may be very great indeed. In a study of the microfauna of Canadian mosses (Fantham & Porter, 1954), 20 samples of *Sphagnum* were examined and the mean numbers of individuals per gram of dry moss were calculated for a few groups (table 2).

Some of these organisms could survive poor conditions for remarkably long periods. Testate amoebae recovered after 5–8 years in dry moss, and the flagellates *Euglena acus* and *Phacus longicaudatus* and the ciliate *Paramecium aurelia* recovered after more than seven years.

Some techniques for investigating the physical and chemical environment of *Sphagnum* are given in Chapter 5.

Table 1. *In* Sphagnum *pH rises in winter in turf (lawn) and hummocks, but not in pools (Rose, 1953)*

	Sphagnum species	Bog Common West Sussex		Thursley Common Surrey	
		Sept. 51	Jan. 52	Sept. 51	March 52
		pH values			
high turf and hummocks	*compactum*	4.2	5.3	–	–
	tenellum	4.4	5.4	4.4	4.8
	capillifolium	4.5	5.7	3.9	4.6
	papillosum	4.5	5.9	4.1	4.9
	magellanicum	–	–	4.2	4.9
low turf	*pulchrum*	4.9	5.9	4.6	4.8
pool	*cuspidatum*	5.8	5.2	4.9	5.0

Table 2. *Number of individuals from different animal groups per gm of dry moss (Fantham & Porter,1945)*

	Protozoa				others	
Sphagnum species	naked amoebae	testate rhizopods	flagellates	ciliates	rotifers	nematodes
papillosum	440	3640	9920	1000	160	120
subsecundum	1344	1712	26672	2224	176	64
palustre	240	3360	5880	2080	120	360
girgensohnii	over 220000				1160	4680

3 Life in *Sphagnum*

Photosynthetic organisms, excluding flagellates

Growing amongst the bog plants, and sometimes on them or in them, are innumerable chlorophyll-bearing organisms including desmids, diatoms, other algae, and cyanobacteria. All can photosynthesise, fixing energy from the sun and eventually passing it on to animals and decomposers. These microscopic green organisms form the base of the food chain for the organisms inhabiting *Sphagnum*. The bog moss plants themselves seem to be inedible except, in minute quantities, to a very few worms and insect larvae.

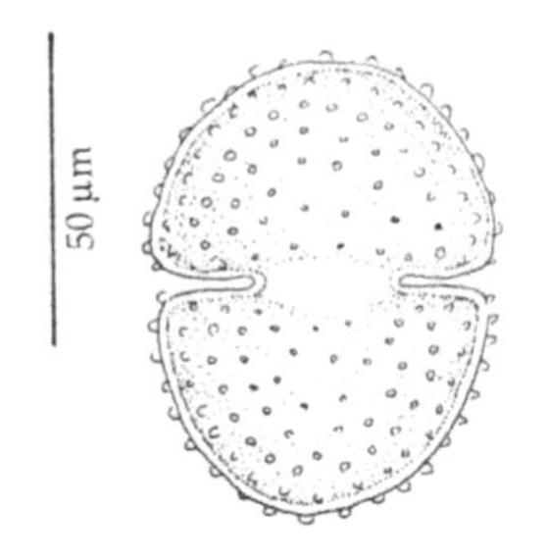

Fig. 8. *Cosmarium tetraophthalmum*

A few drops of the water squeezed from moist or wet *Sphagnum* usually contain a great variety of photosynthetic organisms containing chlorophyll and in some cases additional pigments (table 3, p. 8). All except the flagellates are dealt with in this section.

Desmids

The vast majority of desmid species are confined to districts of acidic rocks and soft water. Most are found in boggy pools and very many live in association with *Sphagnum*. There may be large numbers of both species and individuals. One possible reason why desmids flourish in bogs and amongst bog mosses is the ability of the moss to increase the acidity of the surrounding water by a system of ion exchange (p. 5). This increases the availability of free carbon dioxide, which helps desmids and other plants to flourish. Many more desmids occur in the pool *Sphagnum*, where acidity and water content are relatively constant, than in the hummocks and lawns.

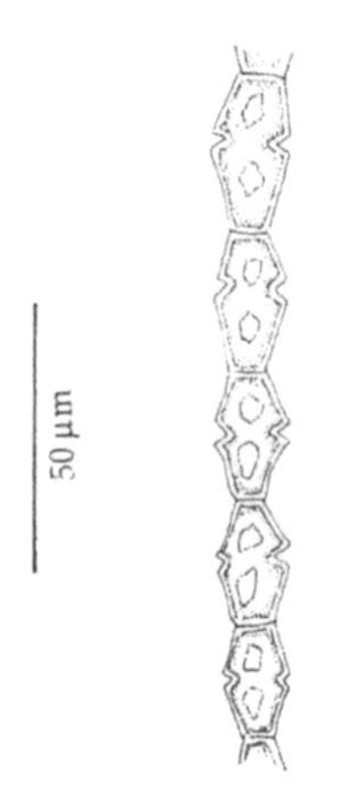

Fig. 9. *Bambusina brebissonii*

Each desmid consists of a cell divided into symmetrical halves or semi-cells. In some genera the halves are more or less divided from one another by a waist or constriction (fig. 8); in others many cells are connected together to form a long filament (fig. 9). The cells range from less than 10 micrometres in diameter to over 1 mm (1000 micrometres) in length. Shape varies widely and desmids of similar shape usually belong to the same genus.

Desmid chloroplasts are rich dark green. They are larger and fewer than those of higher plant cells and they vary in structure and arrangement from one genus to another. In the most highly evolved desmids the chloroplasts are close to the cell wall and their surface is increased by ridges, bumps and extensions round the margin.

Desmids glide slowly. It is still not clear how they do it. Protoplasmic streaming may also be observed within the cell and, in some species of *Closterium*, crystals in well-defined terminal vacuoles (pl. 2.3) are in a state of constant movement.

Table 3. *Some of the organisms found in* Sphagnum *and mentioned in the text (after Barnes 1984, and Lee, Hutner & Bovee, 1985)*

Common name	Taxonomic group	Examples
	Kingdom Monera	
cyanobacteria (blue-green algae)	cyanobacteria	*Anabaena*
other bacteria	aerobic nitrogen-fixing bacteria	
protists	Kingdom Protista	
diatoms	Phylum Bacillariophyta	*Eunotia*
green algae	Phylum Chlorophyta	*Spirogyra*
desmids	Order Desmidiales	*Mesotaenium*
protozoa	Subkingdom Protozoa	
	Phylum Sarcomastigophora	
flagellates	Subphylum Mastigophora	
plant-like flagellates	Class Phytomastigophorea	
yellow-green flagellates	Order Chrysomonadida	*Myxochloris*
dinoflagellates	Order Dinoflagellida	*Gymnodinium*
euglenoids, green and colourless	Order Euglenida	*Euglena*
other green flagellates	Order Volvocida	*Chlamydomonas*
animal-like flagellates	Class Zoomastigophorea	*Bodo*
amoebae	Subphylum Sarcodina	
rhizopods	Superclass Rhizopoda	
naked amoebae	Subclass Gymnamoebia	*Amoeba*
testate rhizopods (shelled amoebae)	Subclass Testacealobosia	*Nebela*
	Superclass Actinopodea	
sun animals (helizoans)	Class Heliozoa	*Actinophrys*
ciliates	Phylum Ciliophora	*Paramecium*
	Kingdom Fungi	
	Phylum Deuteromycotina	
	Class Hyphomycetes	*Helicosporium*
metazoans (many-celled animals)	Kingdom Animalia	
rotifers	Phylum Rotifera	*Habrotrocha*
nematodes (round worms)	Phylum Nematoda	*Criconemoides*
flatworms	Phylum Platyhelminthes	*Stenostoma*
gastrotrichs	Phylum Gastrotricha	*Chaetonotus*
segmented worms	Phylum Annelida	*Cognettia*
crustaceans	Phylum Crustacea	
	Class Branchiopoda	
waterfleas	Order Cladocera	*Chydorus*
ostracods	Class Ostracoda	*Iliocypris*
copepods	Class Copepoda	*Bryocamptus*
	Phylum Chelicerata	
arachnids	Class Arachnida	
spiders	Order Aranea	*Pirata*
mites	Order Acariformes	*Hydrozetes*
tardigrades (water bears)	Phylum Tardigrada	*Hypsibius*
	Phylum Uniramia	
insects	Subphylum Hexapoda	*Gyrinus*

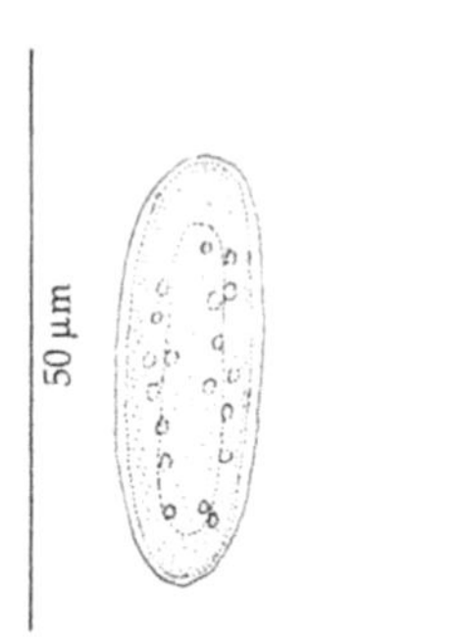

Fig. 10. *Mesotaenium endlichearum*

Desmids are sometimes very abundant. Pentecost (1982) estimated the numbers of *Mesotaenium endlichearum* (fig. 10) as ranging from 300 to 230,000 cells per cm² over 14 different collection sites on Malham Tarn North Fen, and there were 600 to 1500 cells per cm² of *Cylindrocystis brebissonii* (fig. 11) in sites at the edge of a bog pool.

Apart from this study, little is known about the abundance of individuals and the numbers of different species of desmids in different *Sphagnum* habitats. For example, *Sphagnum* species like *S. auriculatum* and *S. recurvum* grow both in moorland bogs and in wet woodlands. Do they support similar desmid floras? One way to investigate this would be to find a *Sphagnum* species growing in both habitats and select, say, ten plants of it from each habitat. They could be sampled by the 'soak and squeeze' technique (p. 54), using a separate labelled dish for each habitat. Sedgewick-Rafter cells are used to count the total numbers of individual desmids and the number of desmid species in 1 ml of water squeezed from plants from each habitat. The diversity of desmid species may be quantified using the Sequential Comparison Index (p. 57). Even less is known about the abundance and diversity of desmids from different *Sphagnum* species in the same habitat. In bog pools, a desmid habitat 'par excellence', bladderwort (*Utricularia*) is an even better microhabitat than *Sphagnum* (Brook, 1981).

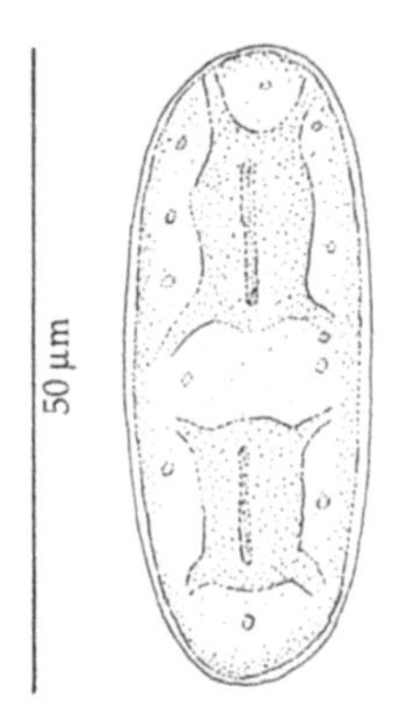

Fig. 11. *Cylindrocystis brebissonii*

Different species of flowering plants in boggy places support very different numbers of desmids, perhaps because of marked differences in plant structure. We have seen how the different microscopic characteristics of *Sphagnum papillosum* and *S. capillifolium*, growing close together on the bog surface, affect the amount of water held in the 'heads' or capitula (Hayward & Clymo, 1982). Does the desmid flora differ in abundance and diversity? *Sphagnum cuspidatum* usually grows submerged in bog pools. How are the desmids distributed throughout the plant? Are they free in the water around the moss plants or do they cluster round the growing point, or on the leaves, stems or branches?

Diatoms

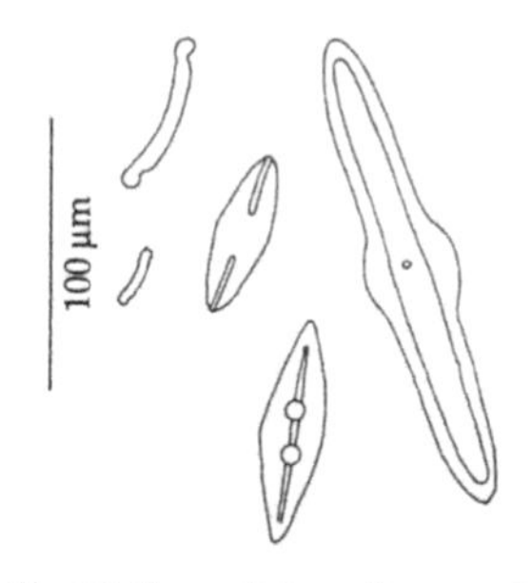

Fig. 12. Some diatom shapes and sizes

Diatoms may be present in very large numbers in *Sphagnum* squeezes. They are often more numerous than desmids, but represent considerably fewer species, especially in very acid bogs. They look yellow-green, because their chlorophyll is masked by golden-brown pigments (xanthophylls).

Diatoms span about the same size range as desmids. A few species are needle-shaped or disc-shaped, but most are rod-shaped: straight, curved or with bulges (fig. 12). Most diatoms are single although they may appear double during division; others form filament-like colonies. Some individual species and genera are restricted to acid water and bogs and these are the ones found amongst *Sphagnum*. *Peronia fibula* grows on *Sphagnum* leaves.

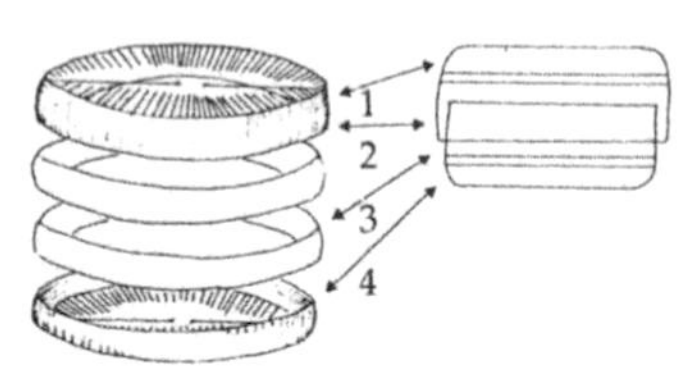

Fig. 13. The diatom frustule or shell in exploded diagram form (after Barber & Haworth, 1981). 1 upper valve, 2 and 3 girdle bands, 4 lower valve

The cell wall of each diatom consists of two halves or valves, one overlapping the other, like the top and bottom of a pill-box (fig. 13). The cell wall is composed of organic matter and silica. When the organic contents of the cell die, the remaining silica shell is transparent, exquisitely sculptured, quite distinctive and easily recognisable in a *Sphagnum* squeeze.

In most freshwater diatoms the pill-box is wider than it is high, so on a microscope slide most individuals lie flat and are seen as it were from the top (or bottom) of the box (in valve view). A diatom which lodges on its edge may look quite different (girdle view). A novice may waste a great deal of time trying to 'key it out ' (fig. 14).

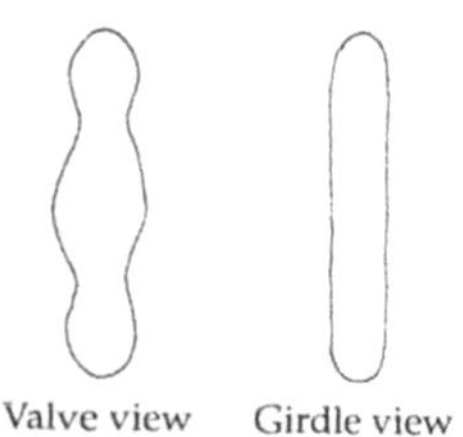

Fig. 14. *Navicula*

Stripes, ribs, beads, pores and other markings on diatom valves make them objects of great beauty under the microscope. They are often used to test the ability of a microscope to resolve fine detail rather than merely to enlarge.

Under the microscope, diatoms are often seen gliding across the field of view. Although the movement resembles that of desmids, different mechanisms have been proposed to account for it.

For diatoms, as for desmids, little is known about the distribution and relative abundance of individuals and species, either in *Sphagnum* growing in different habitats or in different species of *Sphagnum*.

Other algae

Of the green algae that live in the film of water around the bog moss plants, some species occur widely elsewhere; a few (fig. 15) are characteristically associated with *Sphagnum*. Close associates include single-celled forms such as the lemon-shaped *Oocystis*, and the larger, spherical *Eresmosphaera viridis*. *Asterococcus superbus* consists of a small group of cells in a mucilaginous envelope. *Chrysostephanosphaera globulifera* is a flattened, disc-like colony of 2–32 non-motile, naked cells regularly arranged in mucilage, and living in partnership (symbiotically) with the algal cells. Filamentous algae are common, especially amongst submerged *Sphagnum*. Those most closely associated with *Sphagnum* include *Geminella minor*, and *Zygogonium ericetorum* which has violet-coloured sap and forms a purplish sheet on bog mosses in dried-out pools.

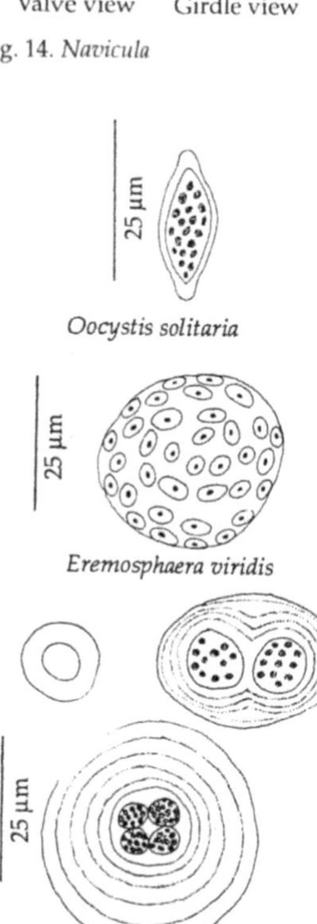

Fig. 15. Some green algae

Cyanobacteria

Formerly known as blue-green algae, these organisms (fig. 16) have now been shown to have more affinities with bacteria than with algae. They have no chloroplasts, nuclei or mitochondria but they can photosynthesise. The chlorophyll, together with other pigments, is distributed throughout the cell, making it appear olive- or blue-green. Many species may contain gas vesicles which act as

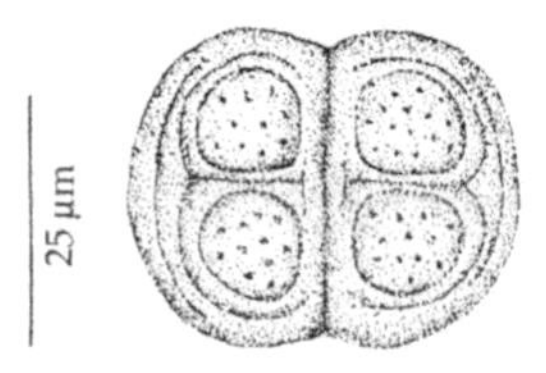

Fig. 16. *Gloeocapsa turgida*

buoyancy regulators. These tiny packets of gas scatter the light, and make the cells appear dark and granular under the microscope. The cells are tiny, averaging only about 5 micrometres in diameter. A few, like *Gloeocapsa turgida* (fig. 16) have larger cells. Like the green algae, cyanobacteria may be solitary, colonial or filamentous. Some of the filaments or trichomes can glide. Some species have specialised cells called heterocysts which are known to be sites of nitrogen fixation.

epiphyte: a plant that lives attached to the surface of another plant

Epiphytes

Some photosynthetic organisms grow attached to the stem, branches or leaves of *Sphagnum*, especially on submerged *S. cuspidatum*. These are called epiphytes. Many of them are filamentous green algae which also occur elsewhere. Others are single-celled green algae, more or less closely tied to *Sphagnum*, including *Octogoniella sphagnicola*, which grows only on the green cells of *Sphagnum* (Pascher, 1930a). Some single-celled epiphytic species have additional photosynthetic pigments. *Perone dimorpha* (Pascher, 1932) is a yellow-green alga (a member of the Xanthophyceae) with a complex life history (pl. 5A). It lives on green cells of the leaf. *Stephanoporos sphagnicola* (fig. 17) and *S. scherffelii* do not undergo the changes of form seen in *Perone*. They were described by Pascher as yellow-brown algae (Chrysophyceae) but in table 3 and the list on p. 47 they are placed in one of the flagellate groups of the protozoa.

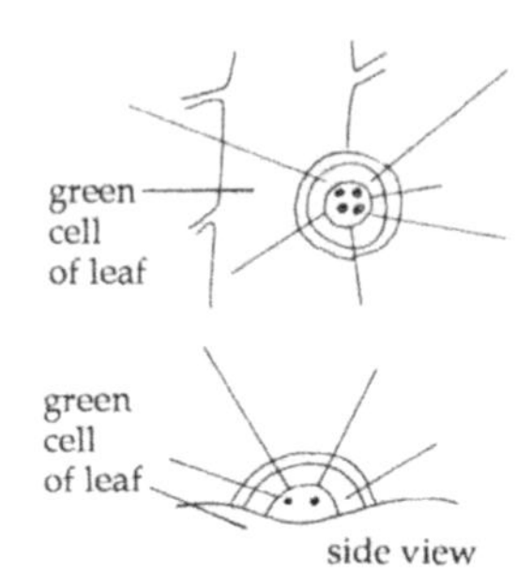

Fig. 17. *Stephanoporos sphagnicola* (after Pascher, 1930).

endophyte: a plant that lives within the structure of another plant, for example inside the hyaline cells of *Sphagnum*

Endophytes

Occasionally a hyaline cell of the leaf may contain a desmid, a diatom or another alga; sometimes older cells are packed with them. Just as water can flow freely in and out through the pores, so can organisms. It should be remembered that these hyaline cells are dead and represent an extension of the environment, rather than part of the living tissue of the plant. Organisms which spend most of their lives inside these cells are known as endophytes. These include *Heliochrysis sphagnicola* (fig. 18) which is closely related to *Stephanoporos* (Pascher, 1940).

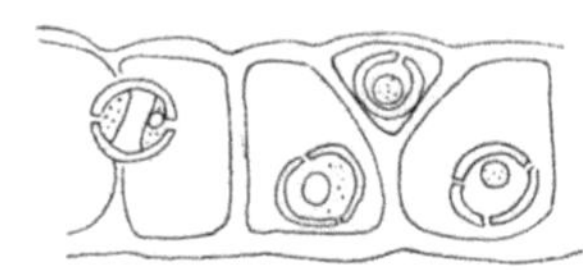

Fig. 18. *Heliochrysis sphagnicola* in hyaline cell (after Pascher, 1940).

Myxochloris sphagnicola possesses green chloroplasts but has a complex life history (fig. 19, pl. 5B); the main stage is amoeboid or plasmodial and the dispersive phases have flagella. It is described in detail by Pascher (1930b) and is very common, although little is known about its distribution. In which species of *Sphagnum* does it live? Is it found in pools, hummocks, 'lawns' and woodland habitats? At what levels of the plant is it found and at what seasons?

A search for *Myxochloris* is likely to reveal some of the less common endophytes. *Chlamydomyxa labyrinthuloides* (Pascher, 1930b) has a large amoeboid body with many green chloroplasts, nuclei, oil drops and crystals of calcium oxalate. Its taxonomic status is problematical (table 3), but it has sometimes been classified with the amoeboid protozoa.

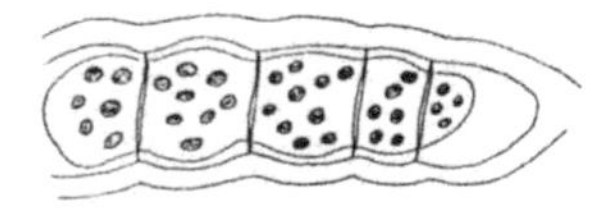

Fig. 19. *Myxochloris sphagnicola* plasmodium with many chloroplasts in hyaline cell.

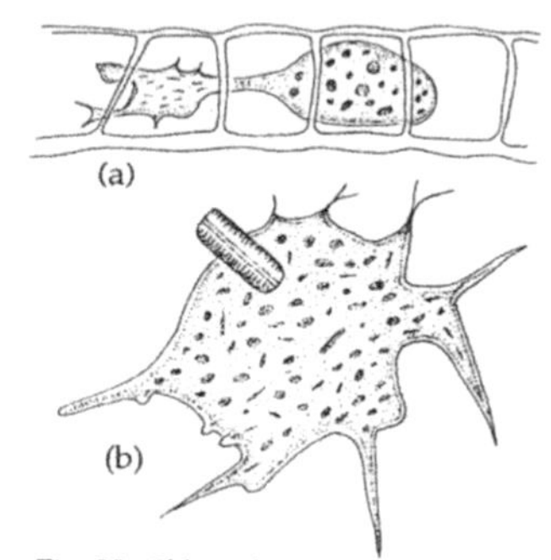

Fig. 20. *Chlamydomyxa labyrinthuloides*
(a) inside hyaline cell
(b) outside cell, capturing a diatom.

For part of its life history it occupies hyaline cells of *Sphagnum* (fig. 20). When it moves outside these, its numerous branching pseudopodia capture large organisms such as desmids and diatoms. It can form cysts similar to those of *Myxochloris*.

The resting spores of *Phyllobium sphagnicola* produce vivid green blotches on old bleached *Sphagnum* leaves (fig. 21). It is a green alga forming filaments which grow through the leaf and produce reproductive organs which release motile gametes. These fuse to form bodies which can infect new cells (Richardson, 1981).

The hyaline cells of the moss are sometimes invaded by the cyanobacterium *Hapalosiphon*, which may become imprisoned within them. Nitrogen fixed by the bacteria may pass to the moss (Stewart, 1966). Species of *Anabaena*, another genus of cyanobacteria, are often found growing from the surface of the leaf or living inside the hyaline cells (fig. 22).

In the mires of Swedish Lapland, cyanobacteria growing epiphytically on *Sphagnum* are the main or only organisms capable of fixing nitrogen, and hence they play an important role in this nitrogen-poor ecosystem (Basilier, Granhall & Stenstrom, 1978). The relationship between *Sphagnum* and its epiphytic and endophytic cyanobacteria merits further investigation.

The water-filled cells of *Sphagnum* form a complex microhabitat which may allow various organisms to survive in a strongly acid environment (Granhall & Hofsten, 1976). The ion exchange mechanism acidifies the water outside the cell, raising the internal pH to a level favourable for cyanobacteria and other bacteria. Perhaps this explains the occurrence of endophytic cyanobacteria at the low external pH values of 4.9 (*Sphagnum riparium*) and 4.2 (*S. lindbergii*). At higher pH the cyanobacteria found in association with *Sphagnum* mosses were all epiphytic; none were endophytes. Fig. 23 illustrates suggested chemical interactions amongst some of these organisms.

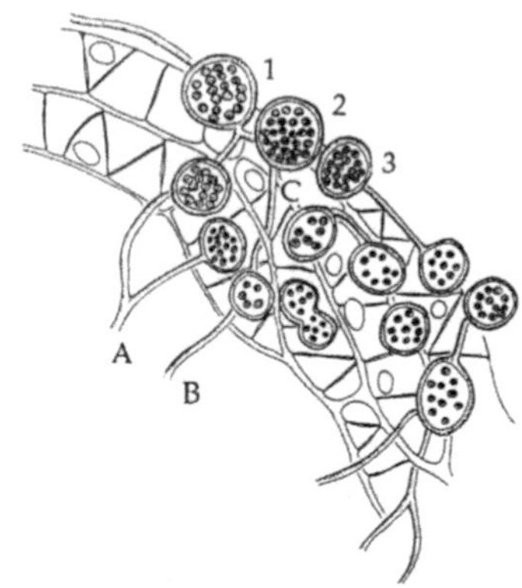

Fig. 21. *Phyllobium sphagnicola* on edge of leaf of *Sphagnum capillifolium*. 1, 2, 3, etc., resting cells attached to vegetative filaments A, B, C, etc., which invade moss cells.

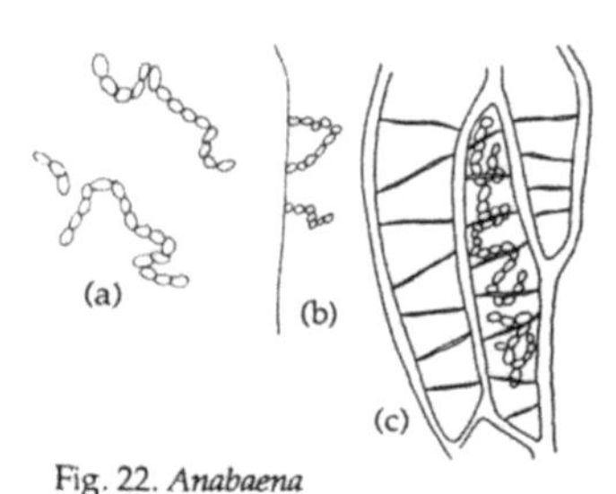

Fig. 22. *Anabaena*
(a) free-living in water film
(b) attached to *Sphagnum* leaf
(c) inside hyaline cell.

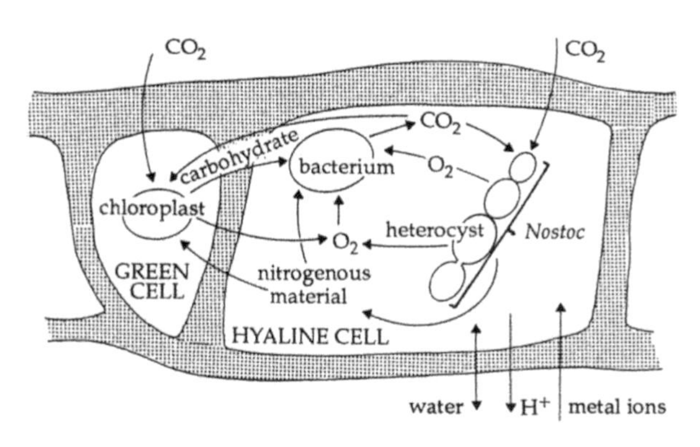

Fig. 23. Hypothetical interactions within leaf cells of *Sphagnum* (modified after Granhall & Hofsten, 1976).

I have found all the epiphytes and endophytes mentioned above, except *Octogoniella*, during examination of material in Southern England, Wales and Scotland. It is certain that other species exist, even perhaps within the living green cells. There remain many discoveries to be made.

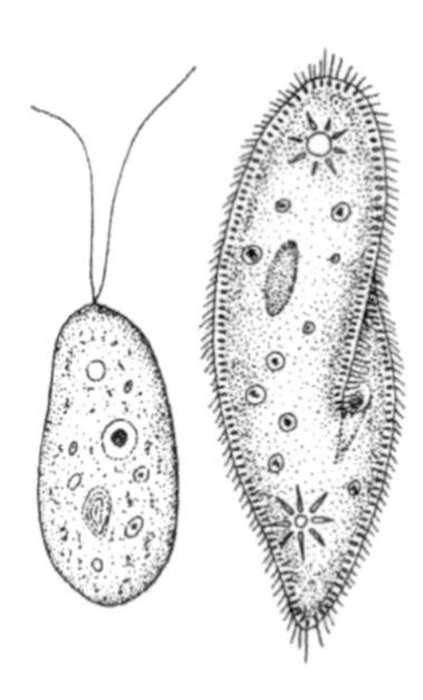

Fig. 24. A flagellate (left) and a ciliate (right).

Flagellates and ciliates

Of the single-celled organisms which swim by means of specialised locomotory processes, flagellates have 1–4 long whip-like flagella and ciliates have numerous fine, short, hair-like cilia (fig. 24).

Flagellates

Flagellates range in size from the tiny *Bodo*, only 5 micrometres long, to species of *Euglena* which can exceed 200 micrometres in length. The majority of species are around 30–50 micrometres long. In some, as in *Platydorina* (pl. 6.7), many individuals are grouped together in a colony. Many species have chlorophyll and can photosynthesise; others are colourless and have different forms of nutrition.

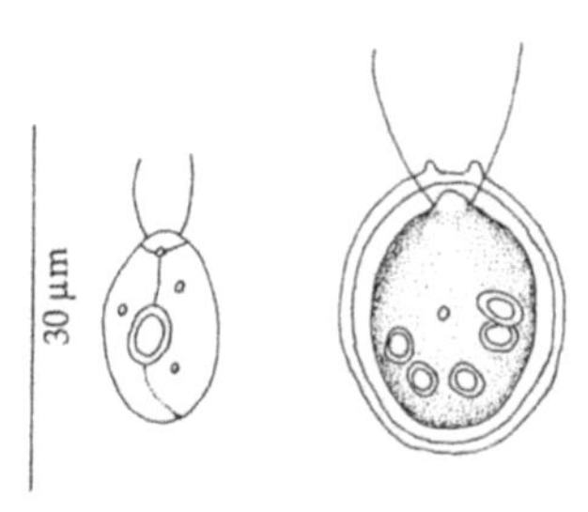

Fig. 25. *Chlamydomonas acidophila* (left) and *C. sphagnicola* (right) (after Pentecost, 1984).

Green forms include *Chlamydomonas*, with rigid cell walls, and the deformable *Euglena*. *Chlamydomonas* species are egg-shaped with two long flagella. *C. acidophila* is common amongst *Sphagnum* at pH 2–6; Pentecost (1982) estimated the numbers of individuals of *Chlamydomonas* (mostly *C. acidophila*) as 50,000 per cm^2 of *Sphagnum* carpet from Malham Tarn North Fen. *C. sphagnicola* is another species found amongst *Sphagnum* (fig. 25).

Euglena species are elongate and carry out squirming or euglenoid movements. There is usually one flagellum. *Euglena mutabilis* is the most characteristic species of acid habitats and can withstand a pH as low as 1.8. It lacks a flagellum and has only two chloroplasts (fig. 26). Pentecost found this species widely distributed in the surface layers of five species of *Sphagnum* in Malham Tarn North Fen, with an estimated 50,000–70,000 per cm^2 of ground surface. Its numbers were positively correlated with moisture content, but not with pH or water hardness. The species was most abundant in the top 2 cm of bog moss and declined rapidly below 4 cm. It was frequently found inside hyaline cells of the leaf, as I have also found in Southern England and Scotland. Pentecost found that he had to squeeze, soak and re-squeeze the moss 200 times to remove all the euglenoids. *Sphagnum papillosum* and *S. palustre* had higher numbers of *Euglena* than *S. capillifolium*, *S. fimbriatum*, *S. squarrosum* and *S. cuspidatum*, perhaps because the first two species have hooded leaves. *Euglena mutabilis*, easily recognised and abundant in *Sphagnum*, is a suitable organism for further research. It is frequently found in cells occupied by rotifers. This association has yet to be described quantitatively and its significance remains unexplored.

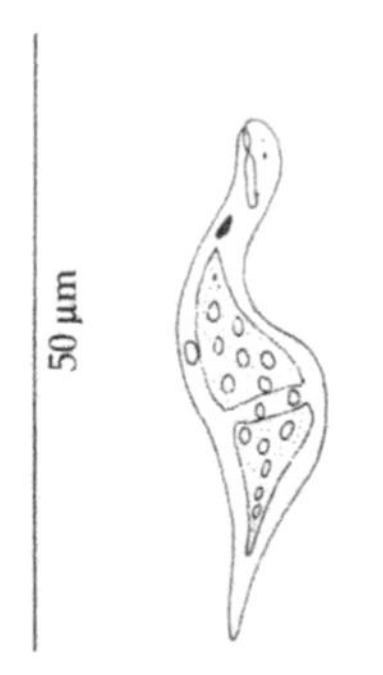

Fig. 26. *Euglena mutabilis* (after Pentecost, 1984).

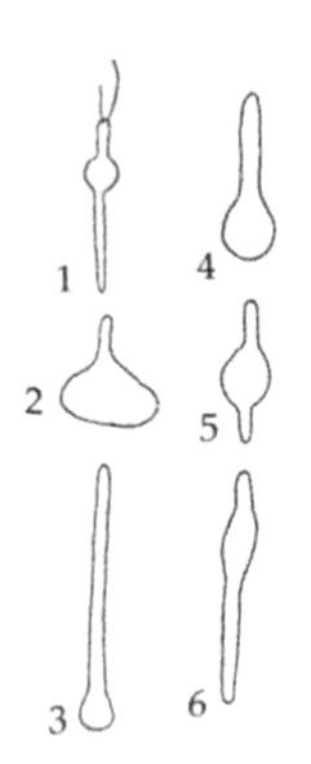

Fig. 27. *Distigma proteus*: diagrams to show successive changes in shape.

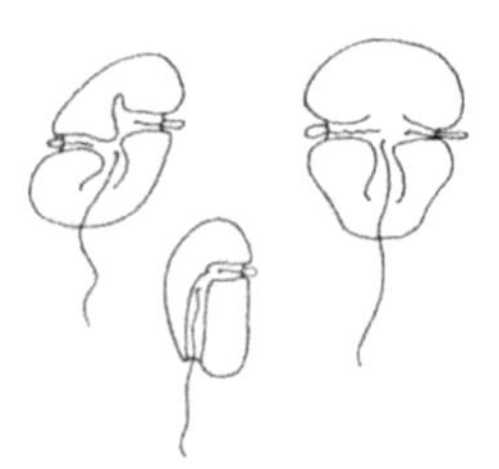

Fig. 28. Some dinoflagellates

Fig. 29. *Lacrymaria*

contracted extended

flask-like container

sphagnum leaf

Fig. 30. *Vaginicola*

One colourless flagellate that deserves special mention is *Distigma proteus* (fig. 27), a bizarre creature, shaped like *Euglena*, packed with colourless granules and with two flagella. It carries out an extreme form of euglenoid movement. Its body changes shape so rapidly as to amaze the first-time observer. The function of these antics is difficult to guess.

In dinoflagellates the two flagella are arranged in characteristic ways. Each lies in a groove, one around the equator of the cell and the other along the length of the cell (fig. 28). Some dinoflagellates have an elegant armoured shell composed of several plates. Most species contain chlorophyll and other pigments; some are colourless.

Dinoflagellates form a large group. Most prefer alkaline conditions, but a few genera favour acidic water bodies and some occur amongst *Sphagnum*. *Hemidinium ochraceum* lives in *Sphagnum*-filled hollows, giving them a yellow-rust colour. Many other genera and species occur in bog pools. Some have lost their flagella and become attached to *Sphagnum* and other bog plants.

Ciliates

Most squeezes of the wet moss yield many ciliates of several species. The smaller ones may be confused with flagellates until viewed at high magnification, when the numerous very fine hair-like cilia should be visible. The cilia beat rhythmically, moving the animal about and producing feeding currents. Some forms, like *Lacrymaria* (fig. 29), have a fairly uniform coating of cilia. In others, cilia are confined to certain parts; for example, in *Vorticella* they form a ring around the mouth. Some have cilia grouped together to form stouter structures called cirri, as in *Oxytricha*. Some large ciliates may look like rotifers, which also move fast. Rotifers, however, are multicellular animals with clearly defined organs and their cilia are limited to a wheel-like ring in front of the body.

Most ciliates are between 40 and 200 micrometres long, but a few species are much longer. The body may be spherical, oval, barrel-shaped, worm-like or bell-shaped (pl. 6). Most move very actively but a few attach themselves to *Sphagnum* leaves by a stalk and some of these secrete a flask-like protective case into which they can retreat (fig. 30). Some enter the hyaline cells of the moss by the pores, apparently feeding on the debris left by former inhabitants.

All ciliates feed on particles. Nearly all hunt for food rather than waiting for it to come to them. Several species look green because, like many other animal species found in *Sphagnum*, they contain symbiotic algae known as zoochlorellae. These may be distinguished from ingested food because they are not enclosed inside food vacuoles. The relationship is mutually beneficial; the algae receive shelter and carbon dioxide and yield oxygen in return.

Grolière (1977), working in the Auvergne in France, found that the ciliate species with zoochlorellae lived in the

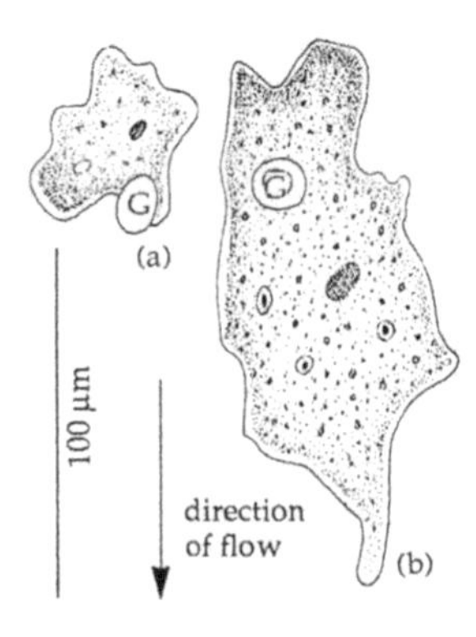

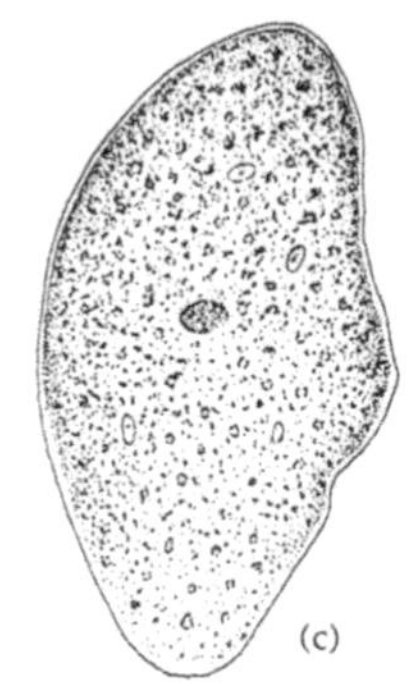

Fig. 31. Naked amoebae
(a) small species
(b) larger species
(c) very dark compact species
G, green epiphyte
captured as food.

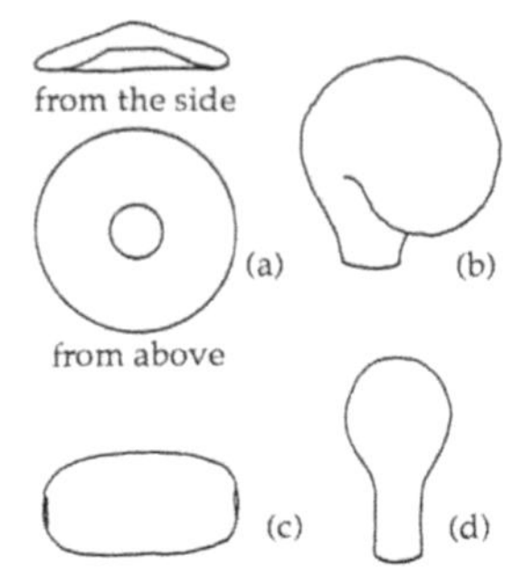

Fig. 32. Tests of various shapes (not to scale)
(a) *Arcella* (saucer shaped)
(b) *Lesquereusia spiralis* (spirally twisted)
(c) *Amphitrema flavum* (a flattened sausage)
(d) *Nebela lageniformis* (flask shaped)

top few centimetres of the bog moss plants with *Cyclidium sphagnetorum* as the dominant species. The deeper, non-green parts of the plants harboured fewer ciliates and these were of different species from those in the zone above. Different species were found in winter and summer. Apart from this study, little is known about the distribution of ciliates in different regions of the *Sphagnum* plant or among different species of *Sphagnum*. This might be investigated using methods described for desmids (p. 9) or testate rhizopods (p. 17).

It is well known that ciliate communities in many different habitats show a succession of different species as environmental conditions change. This possibility has not yet been explored in *Sphagnum* but it could be investigated easily in both field and laboratory. Samples from bog pools could be collected in different seasons and examined for abundance and diversity (p. 56). In the laboratory, samples kept in loosely tied plastic bags in different conditions of temperature, moisture content and light intensity could be sampled at regular intervals. Problems of identification can be overcome by concentrating on easily recognisable genera, or by giving each species a temporary nickname and making a series of careful sketches and drawings (with a scale line) to enable each to be identified, at least to genus.

Naked and case-bearing amoebae and sun animals

Several different kinds of amoebae may be found in *Sphagnum* squeezes. Some are dark and granular, gliding slowly in one direction; others are bright, transparent creatures which extend numerous fluid lobes or pseudopodia (fig. 31). Large naked amoebae are sometimes very numerous in submerged *Sphagnum;* Rogerson (1982) found more than 15 per cm^2 in August.

Green 'amoebae', which live in the leaf cells of bog mosses and change into something quite different at another stage in their life history, are discussed on p.11–12.

Case-bearing amoebae or testate rhizopods

Naked amoebae are so called to distinguish them from those that build themselves cases. It is these case-bearing amoebae for which *Sphagnum* is justly famous.

The case is known as a test, giving these animals their group name of Testacea or testate rhizopods. Because their tests are so distinctive, the species are easy to recognise and identify. They are usually abundant and provide excellent material for small-scale studies of distribution in relation to environmental factors.

The tests range from about 16 to 400 micrometres in length; most are around 100 micrometres long. They may be saucer-, basin-, sausage-, or more often, flask-shaped (fig. 32). A test may consist of transparent secretion, regular or irregular plates or scales, or more or less neatly arranged

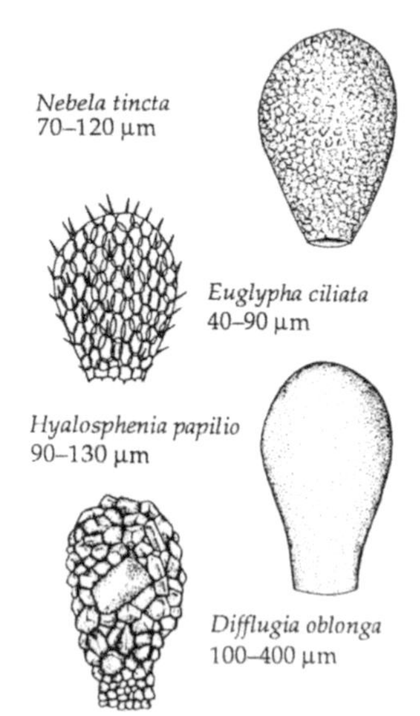

Fig. 33. Four species of testate rhizopods

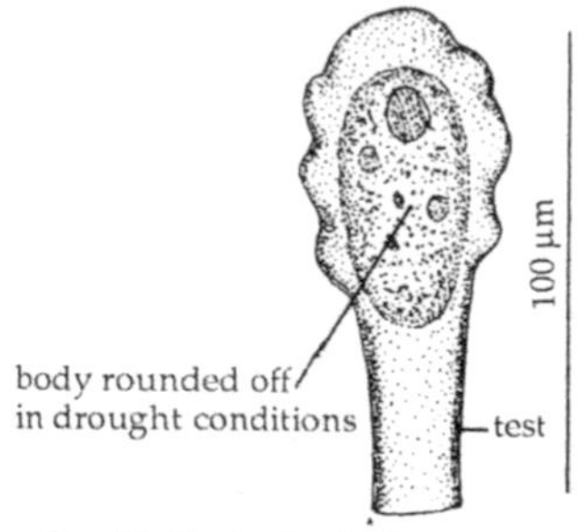

Fig. 34. *Hyalosphenia elegans*

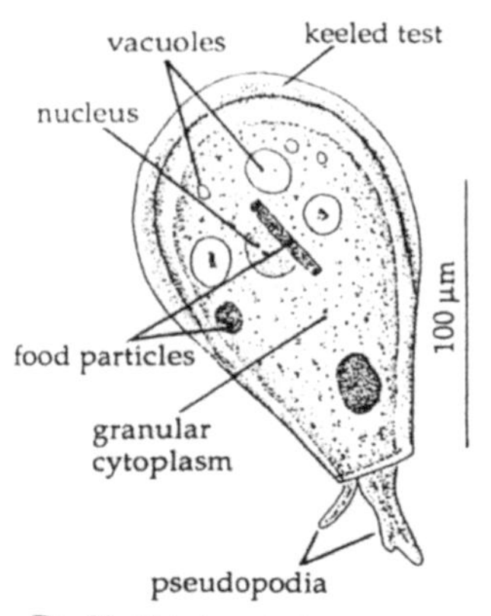

Fig. 35. *Nebela carinata*

mineral or organic particles selected from the immediate surroundings (fig. 33). Some tests are colourless; others are amber, brown, violet or even rose-coloured. The test has a mouth or in some cases two mouths from which the living animal can extend blunt or fine branched pseudopodia. These can engulf food particles.

When searching amongst *Sphagnum*, one often finds empty tests as well as encysted animals in which the contents have rounded off (fig. 34) and in which the test is sometimes sealed with a plug. In the granular cytoplasm of the living animals, food items such as diatoms and green algae may be seen inside vacuoles, undergoing digestion (fig. 35). The forms living in the well-lit regions of the upper few centimetres of the plants include several transparent-shelled species which house green algae called zoochlorellae (pl. 3.5, 4.2). These manufacture food by photosynthesis and the amoebae are thought to benefit from the products (see p. 14).

Case-bearing amoebae are widespread in soil and fresh water, but the species that inhabit *Sphagnum* are particularly diverse, delicate and beautiful. Some are confined to this habitat, including *Amphitrema flavum*, *A. wrightianum* and *A. stenostoma*, *Hyalosphenia papilio* and *H. elegans*. Of those that live in *Sphagnum*, some live amongst mosses submerged in the water of bog pools, while others live in moist plants or in drier situations amongst the hummock-forming species. Woodland bog mosses often contain a large number of species and individuals.

The concentration of oxygen is probably very high by day in the top few centimetres of the moss plants where the living animals are found. Amongst the dying moss lower down there must be very little free oxygen. Here there are only empty cases. Some species of case-bearing amoebae appear to tolerate very acid conditions.

Insignificant as these animals may seem in the scheme of nature, their beauty and variety make them worthy of greater attention than they have received in the past. Corbet (1973) summarises results of earlier workers and draws attention to many questions which remain to be answered.

How are different species distributed within the microhabitat? Vertical distribution on a single moss plant may be studied by dividing the plant into top, middle and lower portions of equal length and using the 'squeeze and soak' technique described on p. 54 to count the numbers of live animals and empty tests of each species. It is useful to mount on a slide 1 cm lengths of the three portions of a neighbouring plant and examine them under a microscope to check this method and to discover the exact position of animals on the plant. Are they in leaf axils, on leaf tips, on leaf surfaces, or even (for small species like *Cryptodifflugia ovalis*) inside the hyaline cells?

How are different species distributed within a moss hummock? This could be investigated for a selected species like *Amphitrema flavum*, which is easy to recognise by its shape, colour and the presence of green zoochlorellae.

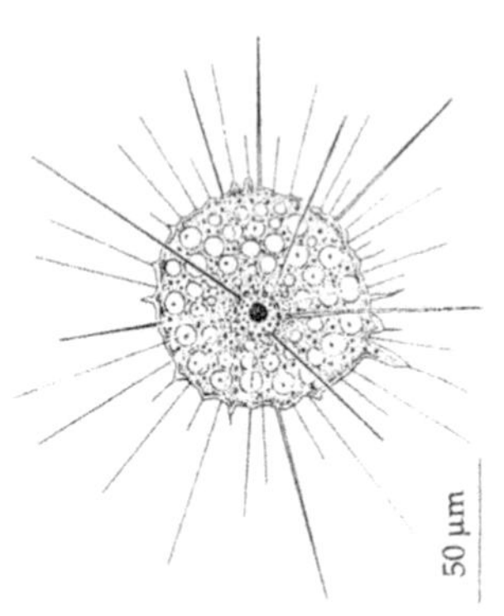

Fig. 36. *Actinophrys sol*

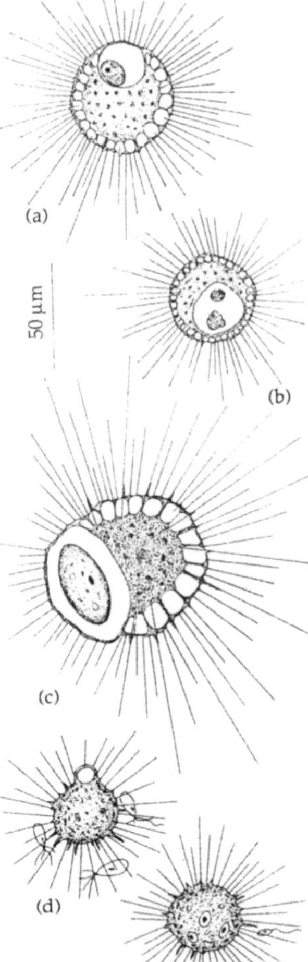

Fig. 37. (a)–(c) *Actinosphaerium* (d), (e) *Acanthocystis*

Horizontal distribution could be explored by examining, say, five points, on a north to south transect across a hummock. Samples could be taken from, say, five depths at each point, to explore the vertical distribution of live animals and empty tests. using the 'squeeze and soak' technique (p. 54). Changes in distribution with time might be revealed by sampling at intervals throughout the day or at different seasons.

How do the animals survive adverse conditions? This question might be approached by estimating the numbers of live animals, encysted forms and empty cases in a habitat after drought, heavy rainfall or extremes of heat or cold. Field observations could be supported by laboratory experiments in which temperature and moisture are controlled. For short periods, animals may be kept on slides with a ring of vaseline supporting and sealing the coverslip, allowing for quick inspection.

Are the animals sensitive to pollution? *Sphagnum* certainly is (Richardson, 1981). If we knew more about the way communities of testate rhizopods respond to pollution, we might be able to read the pollution history of a bog by examining the earlier communities preserved as empty tests at different depths in the peat.

Sun animals or heliozoans

About 20 different species of heliozoans live amongst *Sphagnum*, mostly in the larger bog pools that are not too acid (pH 5–5.6). Smaller numbers of both individuals and species are found amongst lawns or carpets of moss. Great numbers of individuals, but few species, inhabit brooks or ditches intersecting the bog. Heliozoans are most numerous in autumn and spring when the oxygen content of the water is low.

A distinctive and common species is *Actinophrys sol* (fig. 36). About 40–50 micrometres in diameter, it has stiffened rod-like pseudopodia, called axopodia, which reach the centre of the cell.

The most spectacular of all the heliozoans is *Actinosphaerium eichhorni*. It is normally 200–300 micrometres in diameter, but some individuals may reach 1000 micrometres. Members of this genus may be recognised by the vacuole-filled outer layer (fig. 37).

The smaller sun animals are easily overlooked. Some are stalked, others are packed with green zoochlorellae and some have a gelatinous envelope with spines or scales.

It is entertaining and instructive to watch sun animals ensnare and capture their food (fig. 37). Do these animals show any dietary selectivity of food items, or do they capture and digest any moving organism that blunders past the sticky axopodia? This might be investigated by comparing the proportions of diatoms, desmids, flagellates and rotifers present in food vacuoles, or seen to be captured and incorporated into food vacuoles, with the proportions

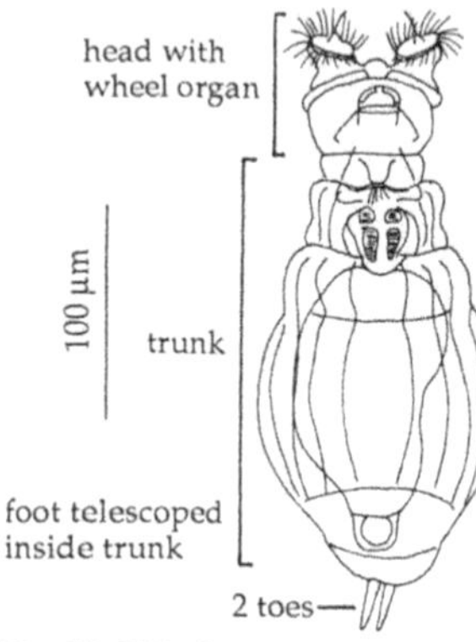

Fig. 38. *Philodina convergens* (after Murray, 1908a).

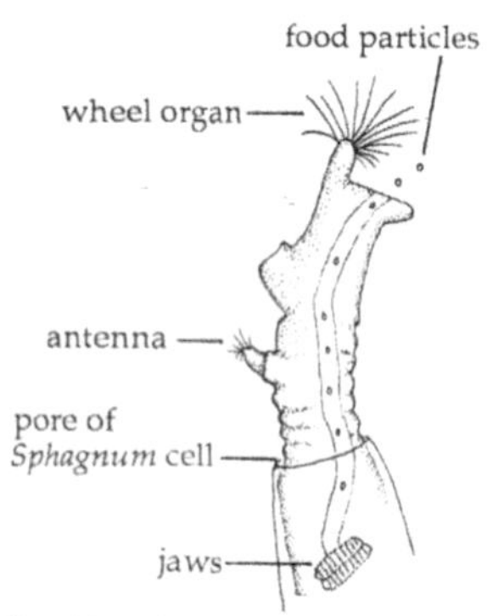

Fig. 39. *Habrotrocha*. The current produced by the wheel organ sweeps particles to the mouth.

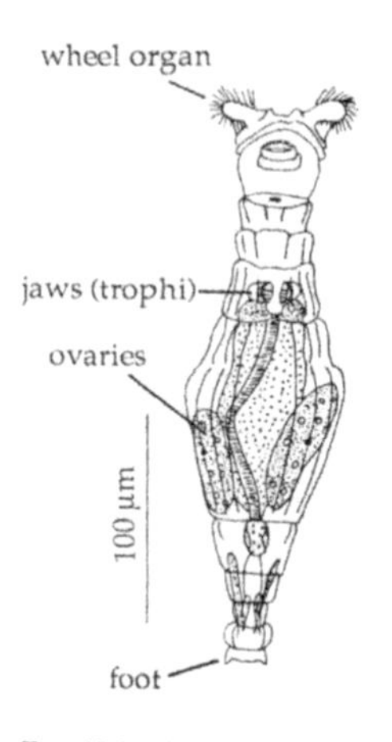

Fig. 40 *Mniobia obtusicornis* (after Murray, 1905).

of these groups available for food in the surrounding water. Sometimes an organism is captured, only to be rejected a few minutes later. Does the probability of rejection depend on the size or species of prey?

Rotifers or wheel animals

Unlike all the organisms discussed so far, rotifers are multicellular animals. They possess true organs composed of specialised groups of cells. A typical rotifer has a head bearing the wheel organ, a trunk, a foot and two toes (fig. 38). These parts are not strictly comparable with the parts of the same name in higher animals although they perform similar functions.

The richly ciliated wheel organ is used in locomotion and food collection. It gives the appearance of a rotating wheel. In the bdelloids, the rotifers most closely associated with *Sphagnum*, and in some other forms, the effect is produced by the rhythmic beating of coarse cilia, forming a whirlpool that draws food particles to the mouth and drives the animal forwards (fig. 39).

In some species the head bears eyespots, usually red, various sensitive papillae and tentacles. Both trunk and foot often appear ringed; by telescoping them the rotifer can greatly shorten the body, changing its shape. The trunk may be soft and unprotected, or it may have various types of 'armour plating', or be enclosed in a tube of some kind. Foot and toe structure are closely related to the mode of life.

Internally, rotifers have digestive, excretory, muscular and reproductive organs. There is also a nervous system with a rudimentary brain. The transparency of the animals allows the internal organs and sometimes food contents and eggs to be seen (fig. 40).

In the simpler organisms described in earlier sections, there is no true distinction between males and females. Rotifers illustrate a most unusual state of affairs. Females are plentiful and large; males are scarce, small, very short-lived and relatively poorly known. In some cases the male is reduced to the bare minimum; it consists of little more than a bladder and a penis.

Rotifers live in a female-dominated society. Amongst the bdelloid rotifers males are quite unknown. Here unfertilised eggs develop into another generation of females. In turn these produce more unfertilised eggs and so on. Amongst the other main group of rotifers, the ploimates, the females of a species all look alike but comprise two distinct types, differing in behaviour. Females of the first type, present for most of the year, produce unfertilised eggs capable of development into females, as in the bdelloids. Females of the second type appear only in unfavourable environmental conditions such as drought or cold. These females each produce a sexual egg. If it remains unfertilised it develops into a male; if its mother was fertilised by a male it develops into a thick-walled resting

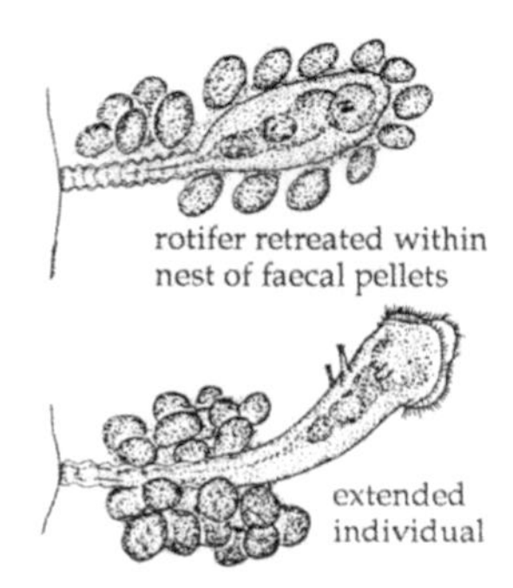

Fig. 41. *Ptygura*, a sessile, nest-making rotifer.

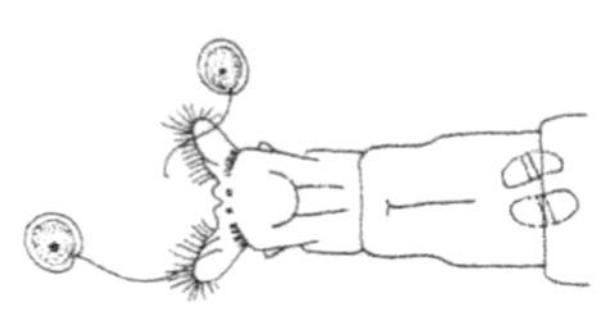

Fig. 42. *Philodina* showing *Trachelomonas* captured by the wheel organ.

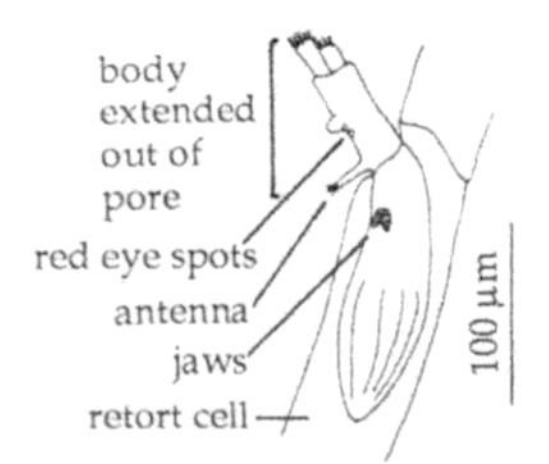

Fig. 43. *Habotrocha roeperi* in retort cell of *Sphagnum* branch cortex.

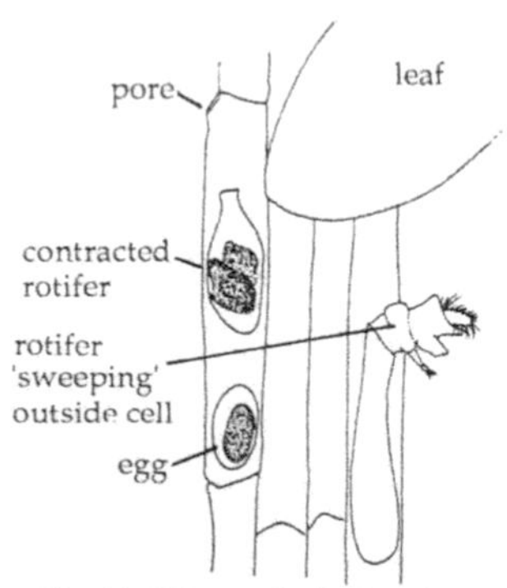

Fig. 44. *Habrotrocha ?reclusa* in retort cells of cortex.

egg which can resist adverse conditions. Resting eggs develop into females.

Some rotifers can produce living young. Eggs are retained in the body until they hatch and the young are born fully developed.

Many kinds of rotifers do not occur amongst bog mosses because the acidity is too great. Some, however, prefer acid conditions and a few live only amongst *Sphagnum*.

The ploimate group of rotifers includes some that swim in water, the planktonic or semi-planktonic forms. Many of these occur in bog pools and in the water film clinging to the moss plants. Numbers of individuals are not great, but they include a diverse array of species, feeding on a great variety of plant and animal material (pl. 7).

Other ploimate rotifers associated with *Sphagnum* have abandoned the swimming habit, fixing themselves to vegetation or some other support. They depend for food on small particles brought their way by currents created by their beating cilia. More vulnerable to predators than active forms, some construct protective tubes and withdraw into these when danger threatens. One such form, *Ptygura velata* (fig. 41), makes a case from its own faecal pellets.

The bdelloids form the other main group of rotifers. They are abundant in *Sphagnum*, creeping and crawling over the moss leaf in a leech-like manner or sheltering between the leaves and branches. Their special niche is the water film clinging to the plants, where they feed on the rich microflora and microfauna. Bdelloids can also swim actively amongst the leaves, seeking fresh pastures as they finish grazing the old ones.

It is amongst the bdelloids that the wheel organ reaches its highest development. With luck you may catch one sitting spinning, with the wheel organ fully extended (fig. 42). The bodies of bdelloids are highly contractile, and so look very different when contracted.

Two species of bdelloids, limited to *Sphagnum*, frequently occupy the outer retort cells of the moss branches. *Habrotrocha roeperi* (fig. 43), the larger of the two, is distinguished by two prominent red eye spots and short cilia on the wheel organ. The other, *Habrotrocha reclusa*, aptly named, has no eye spots and has longer cilia. Each lives within a single retort cell, extending its head through a pore and sweeping in the water film for food particles with the cilia of its wheel organ. It withdraws when danger threatens. One or two eggs are laid in the cell and develop there, presumably forcing the parent to search for an empty cell to colonise (fig. 44).

These two species are very common and occur in several species of *Sphagnum* from both bogland and woodland habitats. Because of their habit they are much easier to study than most of the other rotifers associated with bog mosses. Little is known about their distribution and abundance. Which species of *Sphagnum* and what type of habitat does each species of rotifer most commonly occupy? At which seasons of the year are the adults and the eggs most abundant? What proportion of cortical cells do they occupy in branches from top, middle and bottom of a particular *Sphagnum* species? Can the presence or absence of these endophytic rotifers be related to the structure of the plant? Unlike most other species of *Sphagnum*, the retort cells of *S. palustre*, *S. papillosum* and *S. magellanicum* all have spiral thickening in the cortical cell walls. I have found *Habrotrocha* to be absent from these species, but I have found unidentified rotifers in the leaf cells. It would be interesting to carry out quantitative work to investigate these matters further.

Fig. 45. Contracted bdelloid at start of anabiosis.

anabiosis: a resting state in which an organism can withstand extremes of temperature and desiccation

Bdelloid rotifers can withstand unfavourable conditions, going into a state of suspended animation known as anabiosis. Some can form a cyst. They secrete around themselves a jelly-like material that quickly hardens, becoming a protective shell. In the great majority of animals, preparation for survival of drought involves drying and shrinking of the body contents. When drying begins the rotifer draws both ends of its body into the central part of the trunk, then puckers the two ends drawing them up like purse strings (fig. 45). The globular mass so formed becomes wrinkled, losing a great deal of water. Volume is reduced to about one third or one quarter. In this form rotifers may remain dry as dust for months or even years. When placed in water the animal quickly absorbs it, once more becoming extended and normally active. Extreme cold is resisted in a similar way. During anabiosis all body activities slow down so much as to defy measurement.

The advantages of such behaviour to animals living in *Sphagnum* are obvious; the moss is subject to extremes of moisture content and temperature. Even the normally submerged forms of moss may dry up completely when the water table is very low. Those forming lawns and hummocks suffer much greater fluctuations, being alternately scorched and frozen during summer and winter. The 'dry as dust' rotifers are easily carried from place to place by winds, air currents and on the feet of birds and mammals.

Anabiosis in rotifers has hardly been studied in relation to *Sphagnum*. Some bog moss known to contain active bdelloids could be allowed to dry until weighings on three consecutive days show no loss in weight. A small portion of the moss could then be wetted and a drop of water at once squeezed from it onto a slide. The time taken for different bdelloid species to become active could be noted. It is likely to be a matter of minutes rather than hours. The dry moss could be kept for different periods of time before testing. Responses to cold could be explored by keeping a sample of wet moss containing active bdelloids in a plastic bag in the freezer for a day and sampling it at 10 minute intervals as it warms towards room temperature to note when the bdelloids become active.

Worms

Occasionally while observing some rare and beautiful desmid or following the antics of a tiny flagellate under the microscope, one sees a tremendous upheaval caused by a worm entering the field of view. These worms are mostly less than 1 mm long and have a beauty of their own, as well as a special role in the ecosystem.

Worms that turn up amongst *Sphagnum* are of three types: flat, round and segmented. Worms are more likely to be found in bog mosses from wet woodland or in samples with moorland peat adhering than in moss from pools, hummocks or lawns in bogland.

Flatworms

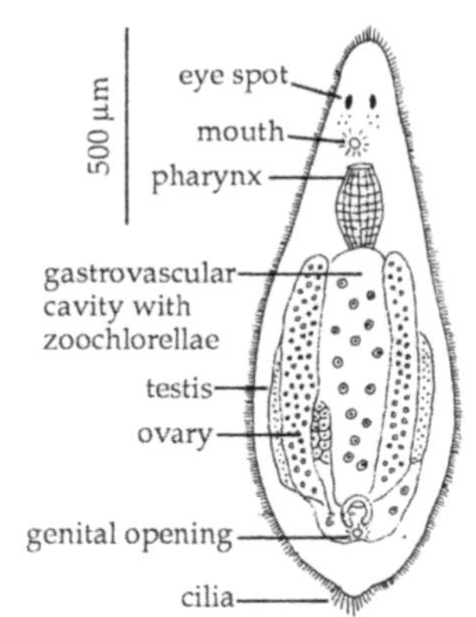

Fig. 46. *Microdalyellia*, a microturbellarian.

Flatworms (Platyhelminthes, Turbellaria) belonging to three groups of microturbellarians are found amongst *Sphagnum*. They resemble large ciliates in shape, in their covering of cilia and in their general habits; but they are flat and have a primitive digestive system, excretory organs and a nervous system which includes sense organs such as eyespots, and a simple brain (fig. 46). Like ciliates they move by gliding. Sometimes they reproduce by simple division and some species carry the process further, producing a chain of daughter individuals. Sexual reproduction occurs in some species, in which case there are true sex organs, each animal being hermaphrodite with both male and female organs.

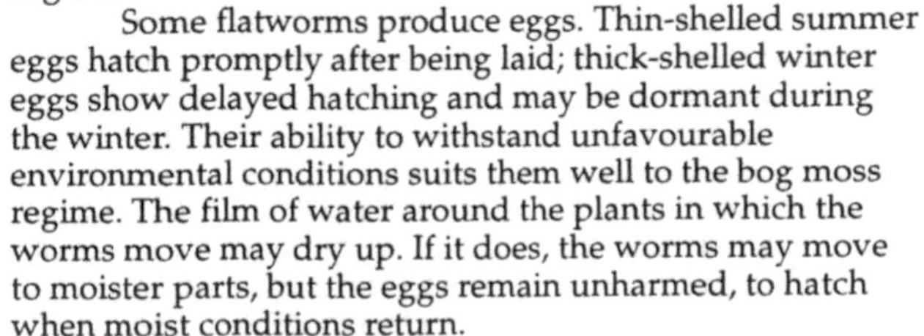

Some flatworms produce eggs. Thin-shelled summer eggs hatch promptly after being laid; thick-shelled winter eggs show delayed hatching and may be dormant during the winter. Their ability to withstand unfavourable environmental conditions suits them well to the bog moss regime. The film of water around the plants in which the worms move may dry up. If it does, the worms may move to moister parts, but the eggs remain unharmed, to hatch when moist conditions return.

Species identification of many microturbellarians is a task for an experienced specialist, because it depends on detailed examination of serial microscopic sections of the body and of the reproductive system of the mature individual. However, quick sketches of living worms to record the key features and size will often allow identification to genus at least.

Roundworms

The free-living roundworms (Nematoda) are much commoner and easier to recognise. It is almost impossible to confuse them with anything else in *Sphagnum*. They are part of a large group of about 30,000 nematode species which occur in soil or fresh water. Only about 30 species have been recorded from *Sphagnum*, worldwide. Woodland bog moss is usually a richer source than bogland *Sphagnum*.

The English name of these worms refers to their round shape in cross section, by which, with their lack of segmentation, they can be distinguished from other worms. The live worms can be recognised by their almost continuous

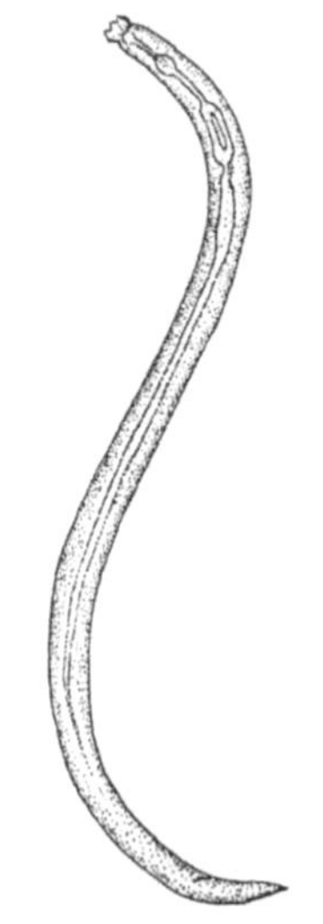

Fig. 47. A typical nematode.

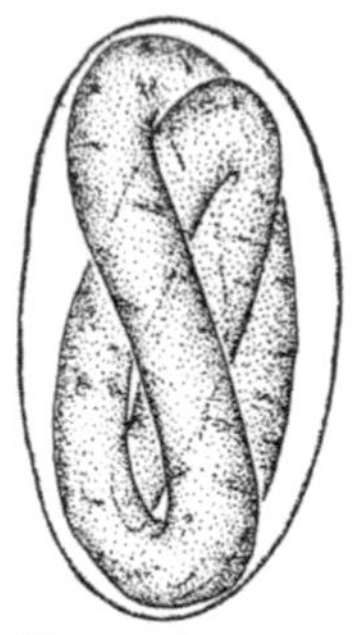

Fig. 48. Nematode egg.

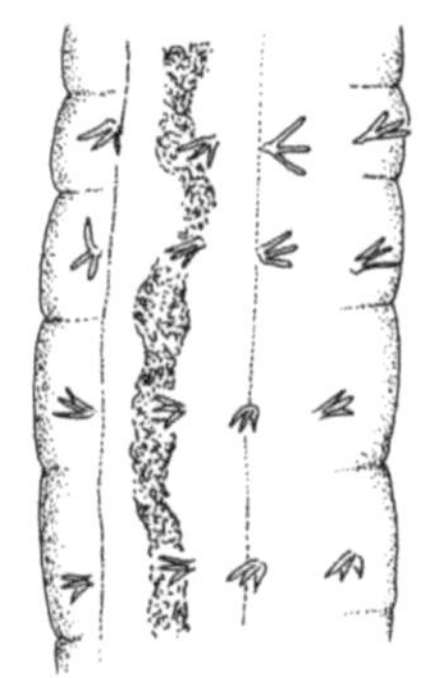

Fig. 49. Part of enchytraeid worm showing bundles of finger-like chaetae.

eel-like thrashing with no change in body diameter and proportions (fig. 47). At the front end, the body tapers slightly, ends abruptly or is bluntly rounded, and there may be a poorly defined 'head'. The hind end usually tapers to a point. These worms are so transparent that their internal organs are clearly visible. The cuticle on the outside of the body may be very finely marked, sculptured or scaly, or it may have prominent grooves around the body, or rows of dots or scale-like folds. Sometimes this marking gives the false appearance of segmentation.

Some nematodes living amongst *Sphagnum* are detritus feeders, ingesting particles of dead organic material. Predatory forms have a variety of modifications for feeding on protozoa and small multicellular animals. Herbivorous species are also found amongst bog mosses but feed on other plant material, fresh *Sphagnum* being apparently inedible except to a few insect larvae.

There are separate males and females but in some species only females are known. The eggs of some nematode species from soil, and presumably from *Sphagnum* (fig. 48), can survive long periods of drought, lack of oxygen and repeated freezing and thawing. The worms themselves can survive unfavourable conditions by adopting a coiled posture, altering their chemical composition by decreasing the percentage of fats, glycogen and glucose and increasing the concentration of glycerine and trehalose. Crowe, Crowe & Chapman (1984) describe the ability of the trehalose molecule to stabilise dry membranes in resting organisms during adverse conditions.

I have frequently found nematode worms coiled up inside hyaline cells of *Sphagnum*, or passing through the pores from cell to cell in the leaves. Eggs are also seen quite often in these cells. It would be interesting to see if this is equally true in different species of *Sphagnum*.

Segmented worms

Much better known to most people are the segmented worms (Annelida, Oligochaeta), the group that includes the common earthworm. The rings round the body (segmentation) and the bristly hairs or chaetae on most segments are characteristic. The species found amongst *Sphagnum* are very small and insignificant compared with the familiar earthworms. They are never as abundant as roundworms, but they can easily be picked out using the petri dish technique (p. 54), and identified at least to genus.

Of the three families of segmented worms found in *Sphagnum*, by far the most important are the Enchytraeidae. Their chaetae resemble a bunch of fingers rather than bristles (fig. 49). Springett (1970) describes six species found in association with bog mosses or the peat formed from them. One of these, *Cognettia sphagnetorum*, shows evidence of daily vertical migration, coming closer to the surface at night. This migration is believed to be a response to the drying out of the plants during the day. *C. sphagnetorum* reproduces by fragmentation and subsequent regeneration and has no drought-resistant cocoon stage. Its distribution indicates adaptation to a cold wet climate.

While the moss plants may provide a suitable moist environment, especially in times of drought, this advantage is counteracted by the poor food value of *Sphagnum*. Only *Sphagnum* stem material has been observed in the gut of *C. sphagnetorum* (Standen & Latter, 1977). Most of the material on which the worm feeds is unpalatable to other animals. However, the waste products (faeces) are colonised by micro-organisms such as fungi and bacteria, and are ingested by animals such as certain protozoa, rotifers and nematodes. The final product of this very incomplete decomposition is peat.

Tardigrades and gastrotrichs

The animals in these two rather obscure groups are perhaps relatively unimportant in the scheme of things. They crop up from time to time in *Sphagnum,* and anyone lucky enough to see them will find them to be both entertaining and instructive.

Tardigrades

Tardigrades are also called water bears. Although they are only a few hundred micrometres long, they resemble land bears in the short stout body, the clawed legs and the slow lumbering gait (fig. 50). They are not very commonly seen in bog mosses; it is a special treat to find one.

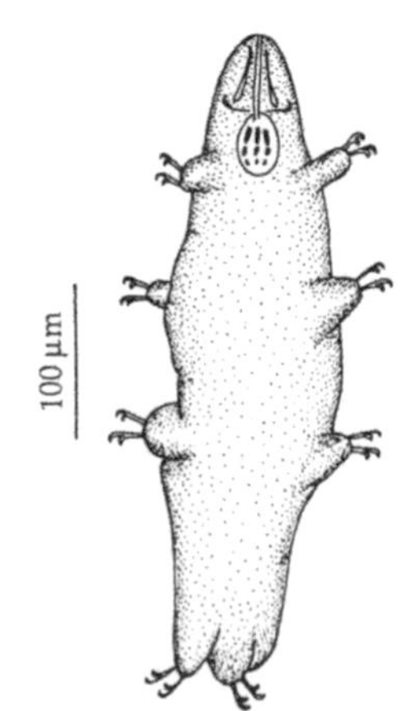

Fig. 50. A tardigrade, *Macrobiotus hufelandii* (after Morgan & King, 1976).

Strictly aquatic animals, they are typically found in droplets and films of water on wet mosses and liverworts and some flowering plants with a rosette arrangement of leaves. They creep about very slowly (hence the group name), using the claws to cling on to plants or debris. The entire body is covered with cuticle which may be variously thickened and marked. A tardigrade has a muscular, sucking pharynx and a pair of long piercing stylets with which it sucks out the contents of plant or animal cells. The water bear moults its skin several times during its life and these remains are sometimes found in mounts or smears of *Sphagnum*.

Tardigrades are masters of inactivity. Like most bdelloid rotifers, they can tide over dry periods. This was first noticed by the early Dutch microscopist, van Leeuwenhoek. To survive dehydration they must be allowed to dry out very slowly and under these conditions the head, hind end and legs are retracted and the whole body becomes more or less rounded into what is known as a 'tun' (Wright, 1989). The animal becomes shrivelled and wrinkled as water is lost from the body. In this state of anabiosis they are highly resistant to the most severe environmental conditions and even to experimental conditions far more extreme than are ever experienced in nature.

Water bears may be aroused from this suspended animation when wetted. Recovery takes a variable time from 4 minutes to several hours. The animal absorbs water, swells and becomes active. This ability must be of great value in bog mosses where the soaking sponge of one season becomes the crisp moss of the next, when most water has evaporated through the pores. Some tardigrades may live as long as 60 years, including anabiotic periods. In the laboratory they can easily be dried and revived ten times or more.

Thin-shelled eggs may possibly be produced without fertilisation during favourable conditions. Thick-shelled fertilised eggs may be produced when environmental conditions deteriorate. In species most likely to be found in *Sphagnum*, eggs are deposited freely, singly or in groups, and are often sticky, spiny or sculptured. In *Hypsibius zetlandicus*, living young are found inside the skin of the mother.

Gastrotrichs

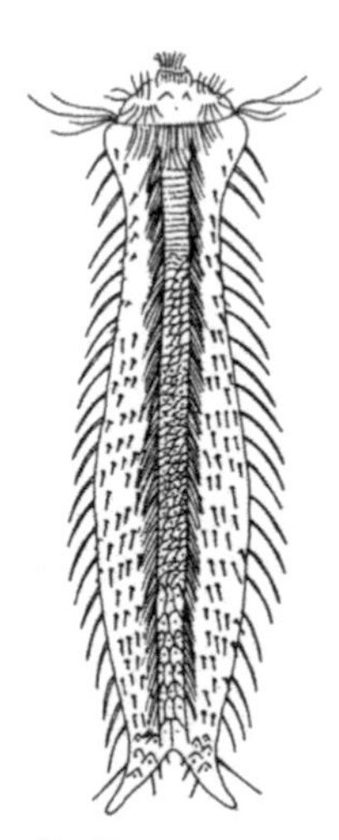

Fig. 51. A gastrotrich, *Chaetonotus maximus* (after Guthrie, 1989).

Gastrotrichs, like tardigrades, once seen are never forgotten or mistaken for any other animals. The group name can be loosely translated as 'hairy stomach' and it is indeed by the beating of long cilia on the lower surface of the body against the substrate, often a bog moss leaf, that the characteristic smooth, graceful movement is achieved. The forked 'tail' is another readily spotted feature of these animals. They range from 100 to 300 micrometres long and have a recognisable head, neck and trunk region. Tufts of longer cilia are usually found on the head (fig. 51). Gastrotrichs are typical of puddles, marshes and wet bogs where there is much decaying material.

Freshwater species are all females which produce viable unfertilised eggs by the process called parthenogenesis. The female-dominated society also seen in rotifers, nematodes and tardigrades is here carried to its final extreme. When environmental conditions become unfavourable larger eggs are produced and these have a heavier shell, surviving desiccation, freezing and unusually high temperatures. Dormancy lasts for a variable period, in some cases as long as 2 years. The adult animals appear to have no mechanism for surviving adverse conditions.

Arthropods

Invertebrates with jointed legs belong to the assemblage previously known as the phylum Arthropoda. Those animals most intimately connected with *Sphagnum*, such as rotifers and testate rhizopods, are small enough to inhabit the film of water in the concavity of a leaf which may be only 300 micrometres in diameter. Very few arthropods are as small as this. Yet minute crustaceans swim in the water surrounding the moss plants, seed mites live

Fig. 52. *Bryocamptus*, a harpacticoid copepod (after Pennak) (about 0.7 mm long).

Fig. 53. Copepod spermatophore.

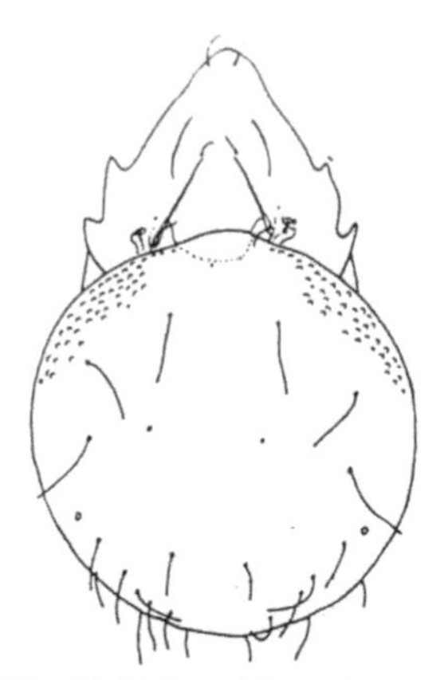

Fig. 54. *Hydrozetes lacustris* (Michael) (450–550 µm; limbs not shown) (drawn by M. Luxton).

between the leaves, spiders run over the moss surface or hide in the stalk layer and several insects have a role within this micro-world.

Crustaceans feed on a wide range of foods found amongst *Sphagnum*, including plant material, protozoa, small metazoa, bacteria and organic debris. Water fleas (Cladocera) and copepods (Copepoda) related to the well-known *Cyclops* are represented by several species associated more or less closely with *Sphagnum*. Amongst these, the cladocerans *Streblocerus serricaudatus* and *Acantholeberis curvirostris* (pl. 8.3) swim in peat pools and the copepods *Moraria sphagnicola* and *Canthocamptus weberi* (fig. 52) are regularly associated with *Sphagnum*. The curious flask-shaped spermatophores (sperm packets) of copepods frequently appear in *Sphagnum* mounts and squeezes (fig. 53). Ostracods, the other crustaceans occasionally reported from *Sphagnum*, are not important members of the microfauna.

Several types of mites may be found amongst bog mosses: oribatid or seed mites of many species, a few species of water mites and some members of the group Mesostigmata. Like spiders, adult mites have four pairs of legs (larval stages have three pairs). Mites differ from spiders in that the body is not divided by a waist.

Water mites (Hydracarina) may be recognised by their bright colours, globular to ovoid shape and their clambering and swimming habits. Conditions are usually too acid for them amongst bog mosses, but it is worth searching for *Hydrovolzia placophora* (pl. 8.1) in cold springs at high altitudes. It is red or orange and favours the axils of *Sphagnum* leaves. Unlike most water mites, it cannot swim (Gledhill, 1960).

Oribatid or seed mites are found in many species of bog moss from different habitats. They are dark and shaped like a tear-drop (pl .4.5–4.8). These mites are more characteristic of soil, but some species are restricted to fresh water. Over 60 species have been recorded from *Sphagnum* in Britain.

Probably the most characteristic genera associated with *Sphagnum* are *Trimalaconothrus*, *Hydrozetes* (fig. 54) and *Limnozetes*. *Hydrozetes lacustris* clambers around the stems and leaves of the moss and *Limnozetes ciliatus* probably has a similar habit. *Trimalaconothrus maior* locates itself in the leaf axils. As many as five species of *Limnozetes* have been collected from a single *Sphagnum* sample by Dr Behan-Pelletier of Canada. *Malaconothrus*, *Mucronothrus* and *Trimalaconothrus* are characteristically associated with wet habitats, but none of the oribatid taxa recorded from Britain is limited to *Sphagnum* (Malcolm Luxton, personal communication).

In a Swedish mire Tarras-Wahlberg (1952) found a distinctive group of species associated with each of three habitats: depressions or pools with *S. cuspidatum*; damp hummocks with mainly *S. magellanicum;* and less damp with *S. fuscum*. The hummocks were the habitat richest in species.

Amongst submerged moss and in the drier parts of the mire he found fewer species but more individuals.

Very little work has been done on the ecology of oribatids in Britain and other countries, partly because identification is so difficult. Even if species cannot be named initially, ecological studies of, for instance, distribution in relation to moisture and *Sphagnum* species can still be undertaken. Sketches or photographs are made of each species which is given a provisional code name or reference number until it can be identified later from properly mounted voucher specimens. A technique for making semi-permanent slide mounts is described on p. 55.

Spiders and insects of many species are common amongst *Sphagnum*. Although some may be small enough to be considered microscopic, they are not considered in detail here. References are given in the key section for interested readers.

4 Identification

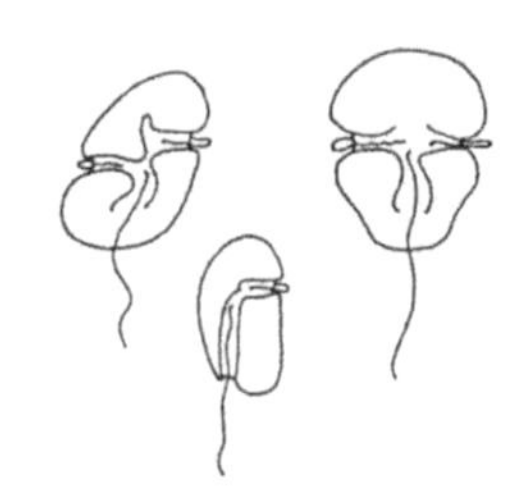

I. 1 Dinoflagellates

The number of plant and animal species which inhabit *Sphagnum* in Britain is very large indeed. A few species are found nowhere else, some are closely associated with the plant but many are equally abundant in other habitats.

With the Keys and Guides in this chapter it should be possible to assign an organism to its main group, or in some cases to genus or even to species. References are given to publications which provide more detailed keys for identification.

In addition, genera and species recorded from *Sphagnum* in Britain are listed on pp. 45–52. Many of these also occur in Europe and North America and some are worldwide in distribution. These lists are by no means exhaustive and some naturalists will be able to add to them both from their knowledge and from the further research which it is hoped that this handbook will stimulate.

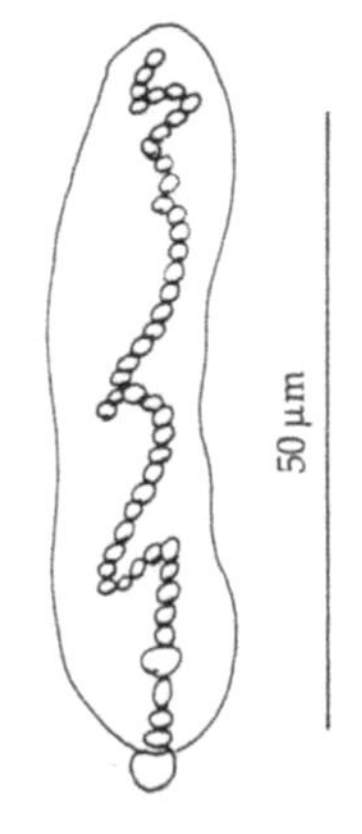

I. 2 Cyanobacteria: *Nostoc paludosum* Kützing in jelly-like case

Key I. Groups of organisms found in *Sphagnum*

At the end this of key are drawings of some miscellaneous items often found in *Sphagnum* which do not key out here (I.28–I.32).

1 With 2 whip-like hairs or flagella, one of which lies around the body; body grooved and sometimes armoured — dinoflagellates (I.1)
– Not having this combination of characters — 2

2 With green, yellow-brown or blue-green pigments — 3
– Without such pigments (except in food vacuoles or in symbiotic algae living inside the body of an animal) — 9

3 Pigment blue-green, diffuse throughout cells — blue-green algae or cyanobacteria (I.2)
– Pigment grass-green, yellow-green or yellow-brown, localised in distinct bodies (plastids) within the cells — 4

4 Pigment yellow-green or brownish — 5
– Pigment grass-green — 6

5 Unicellular, capsule-like, with siliceous (glass-like) walls ornamented by very fine markings; often found empty and colourless — diatoms (Guide I)
– Not like this; usually motile unicells or colonies — golden-brown algae (I.3)

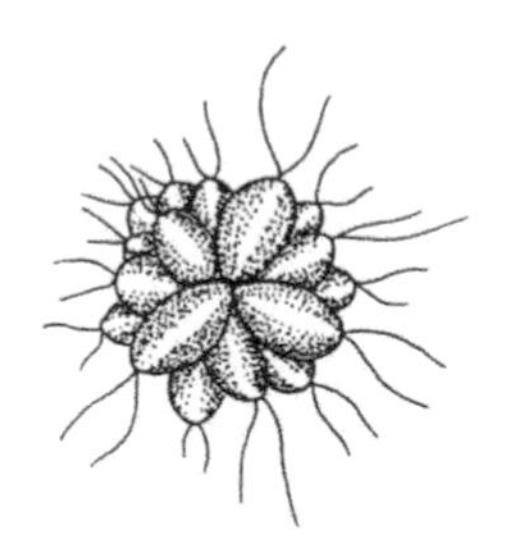

I. 3 *Synura*, colony of golden-brown algae

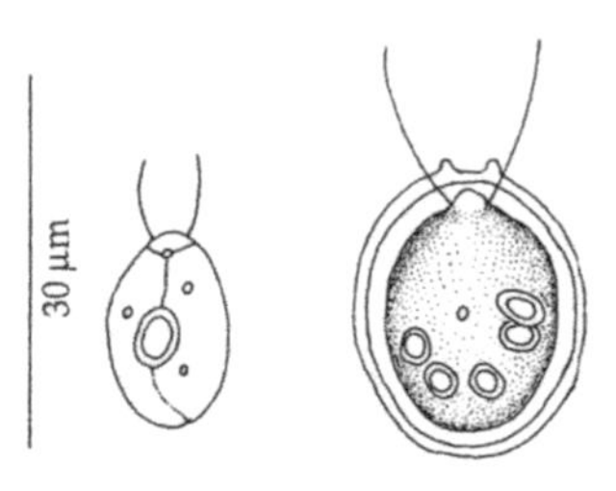

I. 4 *Chlamydomonas acidophila* (left), *C. sphagnicola* (right)

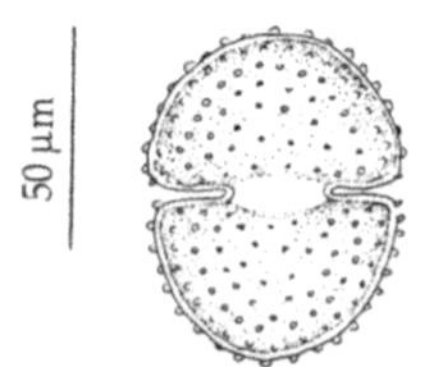

I. 5 *Cosmarium tetraophthalmum*

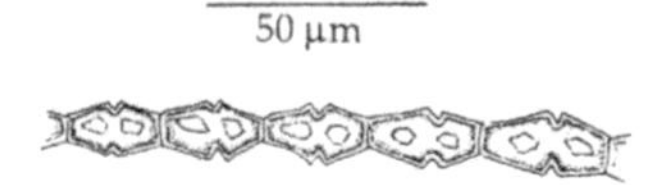

I. 6 *Bambusina brebissonii*

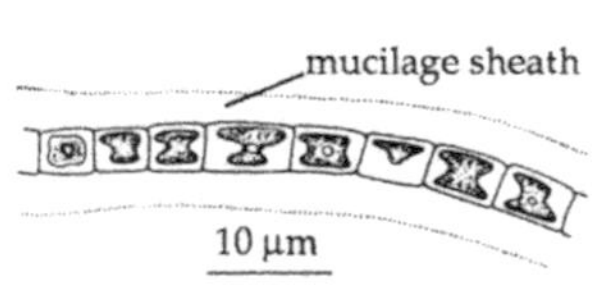

I. 7 *Geminella minor* (Naegeli) Heering (after Pentecost, 1984)

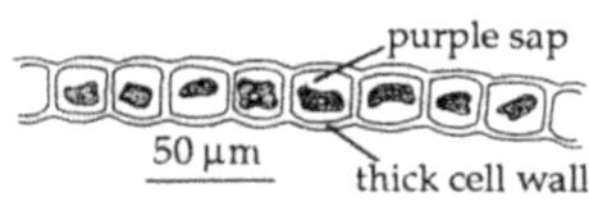

I. 8 *Zygogonium ericetorum* Kützing (after Pentecost, 1984)

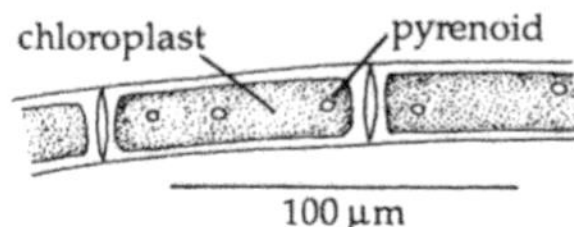

I. 9 Part of *Mougeotia* filament

6 Moves, often spirally, with 1, 2, or 4 beating flagella (fine whip-like structures, sometimes hard to see) (I.4) — plant-like flagellates (Key II)

– Flagella absent; immobile or capable of smooth gliding only — 7

7 Each cell with a rigid cell wall which is often ornamented and consists of 2 identical mirror-image halves (I.5, I.6) — desmids (Key III)

– Not consisting of 2 mirror-image halves — 8

8 Cells arranged in rows forming branched or unbranched threads or filaments — filamentous green algae (I.7–I.9)

Only a few examples are illustrated. Other genera which may be found are listed on p. 46. Some filamentous desmids (I.6) (Key III) may also key out here.

– Cells not arranged in filaments — other green algae

Many species may be found (I.10) (list p. 46). For further identification of green algae see Belcher & Swale (1976, 1979), Pentecost (1984) and Prescott (1951).

9 Very small particles (less than 5 µm) with no internal structure visible under high power of compound microscope; may be spherical, rod-shaped or spiral; often inside hyaline cells of *Sphagnum* — bacteria (I.11)

– Larger, with visible structure or contents — 10

10 Branched filament — brown or colourless fungi (I.12)

Many species of microscopic fungi invade dying *Sphagnum* (see Ingold, 1975). Larger fungi live in hummocks and carpets (Watling, 1973).

– Body not filamentous, but sometimes with branched or unbranched extensions — 11

11 Moves — 13

– Does not move — 12

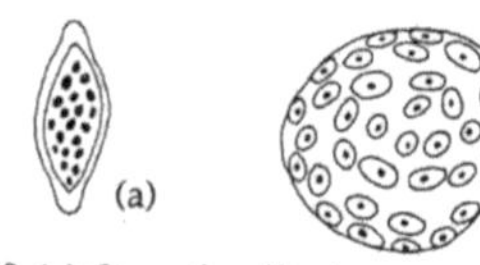

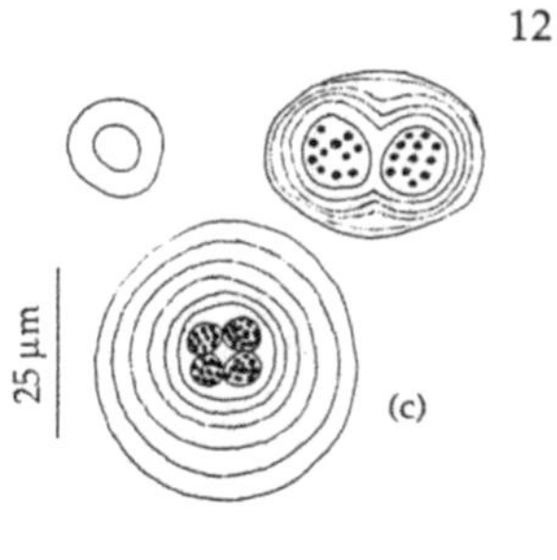

I. 10 (a) *Oocystis solitaria*
(b) *Eremosphaera viridis*
(c) *Asterococcus superbus*

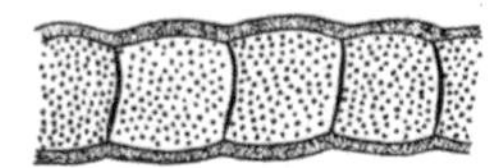

I. 11 Bacteria in hyaline cell of *Sphagnum*

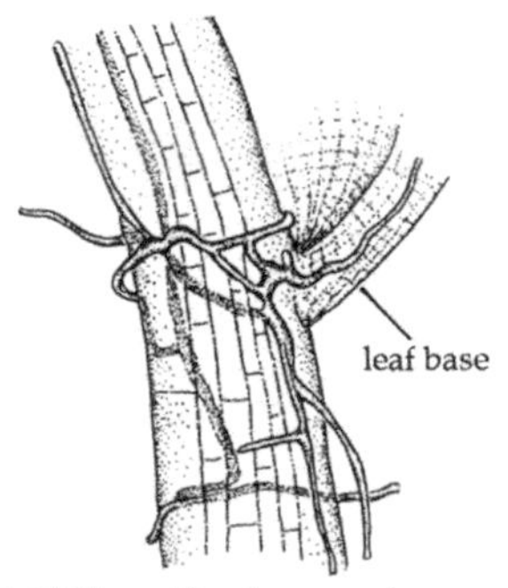

I. 12 Fungal hyphae growing in and around *Sphagnum* leaf and branch

12 Enclosed in a case or test

Perhaps an encysted testate rhizopod (see illustrations in pl. 3 and Key IV) or a contracted rotifer (e.g. fig. 45) or peritrich ciliate (e.g. fig 30). See also I.28–I.32.

– Not enclosed in a case or test

It may be impossible to identify an organism that shows neither characteristic movements nor a case, except by comparison with active individuals from the same sample.

13 Moves by blunt or finely-pointed, branched or unbranched extensions of the cytoplasm called pseudopodia 14

– Moves in some other way 16

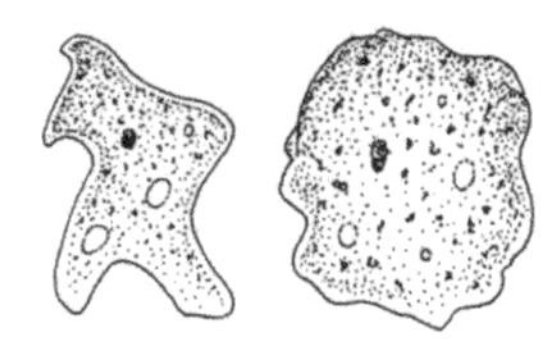

I. 13 Naked amoebae

14 Not enclosed in a case or test 15

– Enclosed in a case or test with a mouth or mouths through which pseudopodia may be extended testate rhizopods (Key IV)

15 Pseudopodia blunt naked amoebae (I.13)

– Pseudopodia long and stiff sun animals (heliozoans) (1.14)

For further identification of naked amoebae see Page (1976) and for heliozoans Cash, Wailes & Hopkinson (1905—21).

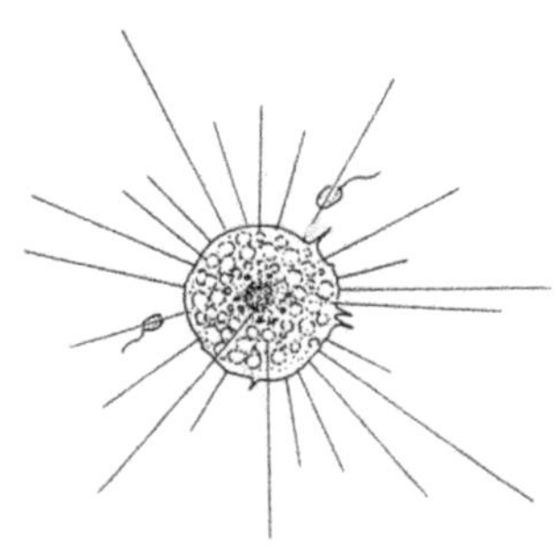

I. 14 Heliozoan

16 Move or create water currents by means of fine hair-like cilia or long whip-like flagella 17

– Move in some other way; without pseudopodia, cilia or flagella 21

17 With one or two flagella, whose beating causes jerky or spiral movement animal-like flagellates (I.15) (Key II)

– With numerous cilia that clothe all or part of the body or form a ring around the mouth 18

18 Unicellular with no true tissues or organs ciliated protozoa (pl. 6)

For descriptions and figures of ciliate genera see Curds (1982) and Curds, Gates & Roberts (1983) and for descriptions of all freshwater species known at that time, including many from *Sphagnum*, see Kahl (1930–35).

– Multicellular; body clothed with cilia or with cilia around mouth region only 19

I. 15 *Bodo*, an animal-like flagellate

19 Cilia around mouth region only (may be retracted if animal disturbed); animal with a delicate or rigid cuticle and a pair of permanently-chomping jaws, the mastax, visible inside the body; sometimes living inside a case or a *Sphagnum* cell rotifers (Key V)

– Body clothed with cilia 20

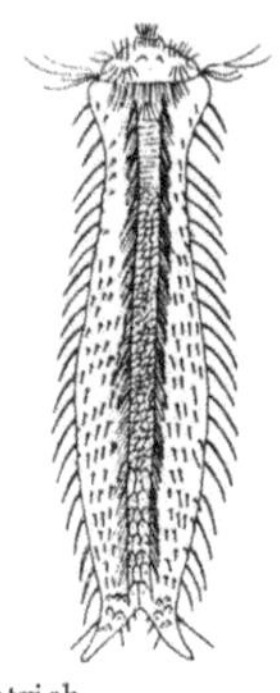

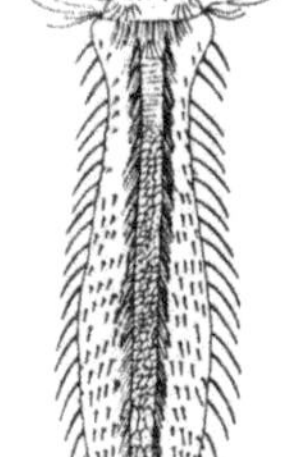

I. 16 Gastrotrich, *Chaetonotus maximus*

20 Small animals, usually less than 300 μm; with forked tail gastrotrichs (I.16)

The taxonomy of this group is not well-known and there is no key to British species. Voigt (1960) gives a key to European species some of which have been recorded from *Sphagnum* in Britain. Thorp & Covich (1991) give a key to genera.

– Larger animals; usually more than 500 μm; without forked tail microturbellarians (I.17)

Young (1970) gives a key some British and Irish species, Rixen (1961) to some German species and Pennak (1978) and Thorp & Covich (1991) to American genera, some of which occur in *Sphagnum* in Britain.

21 Body slender and worm-like 22

– Body not slender and worm-like 24

22 Body unsegmented (some show false segmentation) and with whip-like movements, like an eel roundworms (nematodes) (I.18)

For further identification see Goodey (1951), Goodey & Goodey (1963), Pennak (1978) and Thorp & Covich (1991).

– Body segmented (marked into ring-like sections) 23

I. 17 *Microdalyellia*, a microturbellarian

23 Body encased in transparent cuticle, with an amber-coloured head capsule bearing bristles and mouthparts fly larvae, most of which are chironomids (I.19)

For key to chironomid larvae see Cranston (1982).

– Body softer, more deformable, with no definite head; groups of stiff bristles or finger-like processes on each side of each segment oligochaete worms (I.20)

Brinkhurst (1971) gives a key to *Aeolosoma* species found in Britain.

24 With 4 pairs of unjointed legs, armed with claws tardigrades (I.21)

For further identification see Morgan & King (1976).

– With jointed legs 25

I. 18 A typical nematode

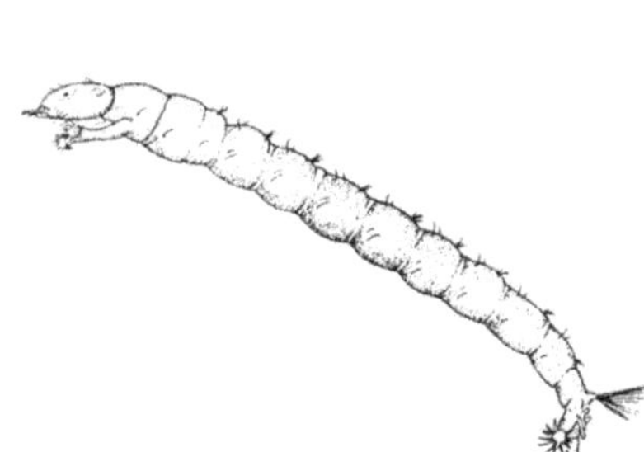

I. 19 A chironomid larva

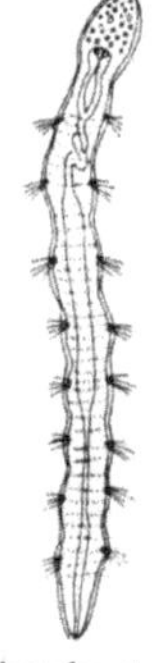

I. 20 Oligochaete worm, *Aeolosoma*

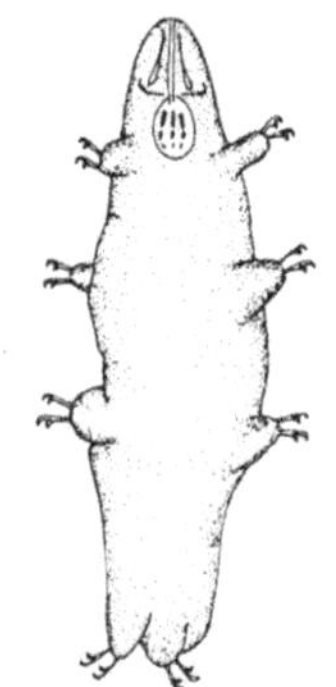

I. 21 Tardigrade, *Macrobiotus hufelandii*

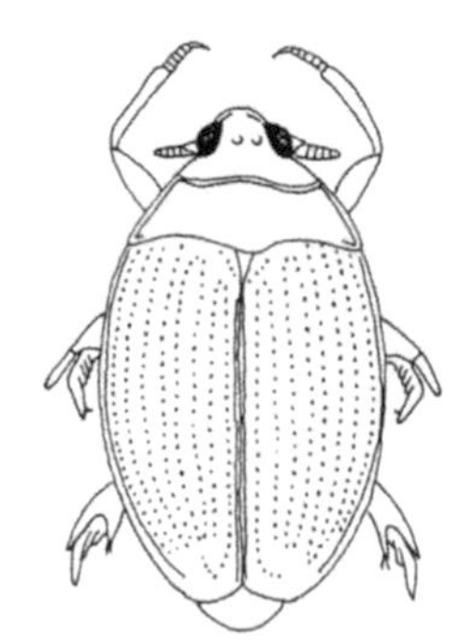

I. 22 Water beetle, *Gyrinus* (about 5 mm long)

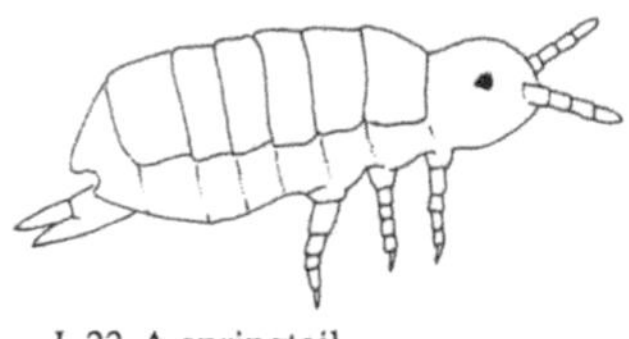

I. 23 A springtail (1–2.5 mm long)

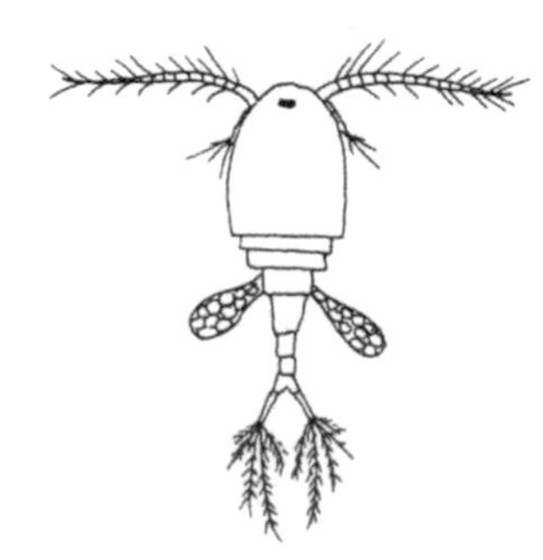

I. 24 A cyclopoid copepod

25 Three pairs of legs — 26
– More than 3 pairs of legs — 27

26 Body divided into 3 regions; head, thorax and abdomen — insects

These include representatives from several orders but the majority are water beetles (I.22). For identification see Friday (1988). Wingless insects include a few species of springtails (I.23). For identification see Fjellberg (1980).

– Body not divided into 3 regions — immature mites

27 Four easily seen pairs of legs — 28
– Legs more than 4 pairs or hard to see and count — 29

28 Body divided into 2 regions — spiders

For identification see Jones (1983) and Jones-Walters (1989).

– Body not divided into 2 regions — adult mites (pls 4 & 8)

These are mainly oribatids. There are no up to date keys to British or European species. Malcolm Luxton's unpublished key forms the basis of the list on p. 5.

29 Body and legs largely enclosed between paired valves — 30
– Body and legs not enclosed; 0.6–3 mm long — copepods (I.24, I.25)

For further identification see Harding & Smith (1974).

30 Valves enclose head and all legs; bean-like creatures which glide steadily along by invisible movements of hidden legs; rare in *Sphagnum*; usually 1–2 mm long — ostracods (I.26)

For identification see Henderson (1990).

– Valves do not enclose head; eyes and antennae visible; may swim jerkily by means of antennae — water fleas (Cladocera) (I.27, pl. 8.3–7)

For identification see Scourfield & Harding (1966) or Amoros (1984).

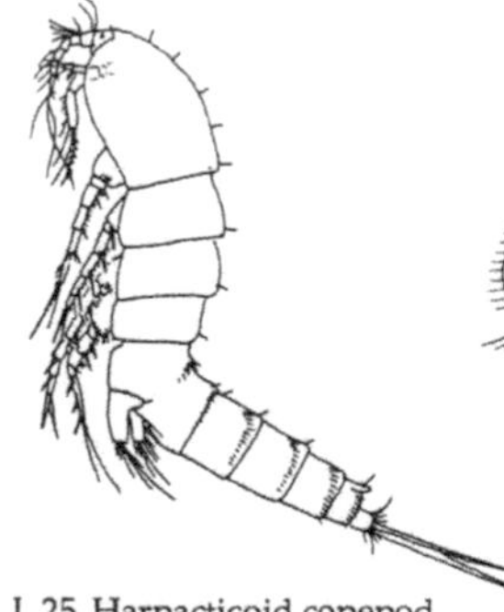

I. 25 Harpacticoid copepod, *Bryocamptus*

I. 26 An ostracod

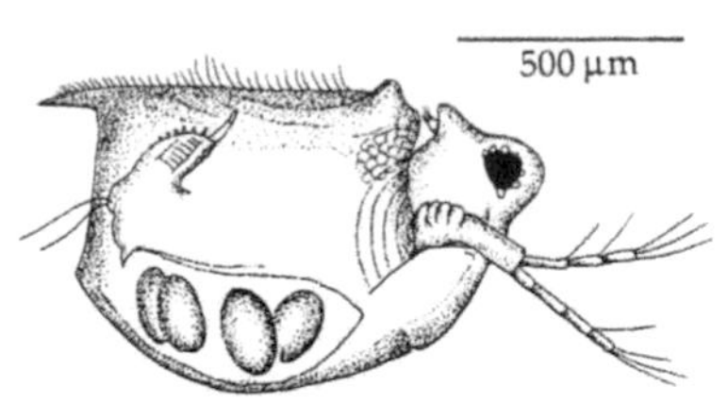

I. 27 Cladoceran, *Scapholeberis mucronata* (after Ward & Whipple, 1918)

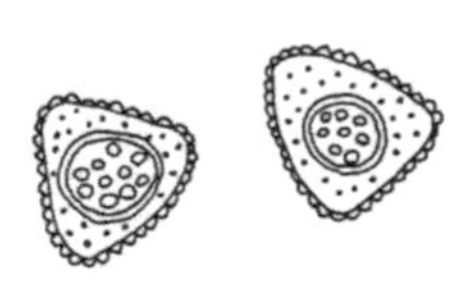

I. 28 *Sphagnum* spores

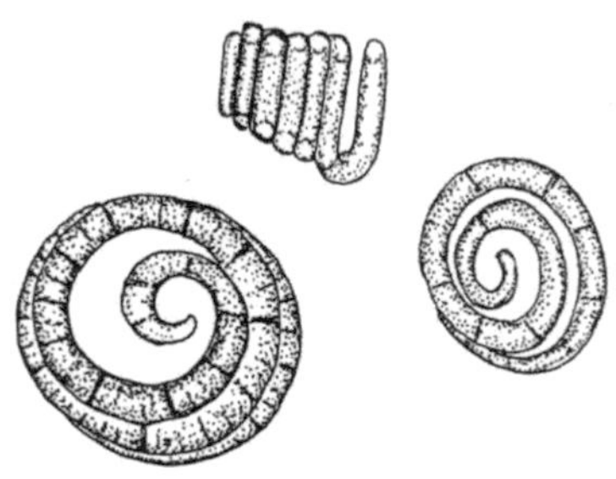

I. 29 Fungal spores of *Helicosporium*

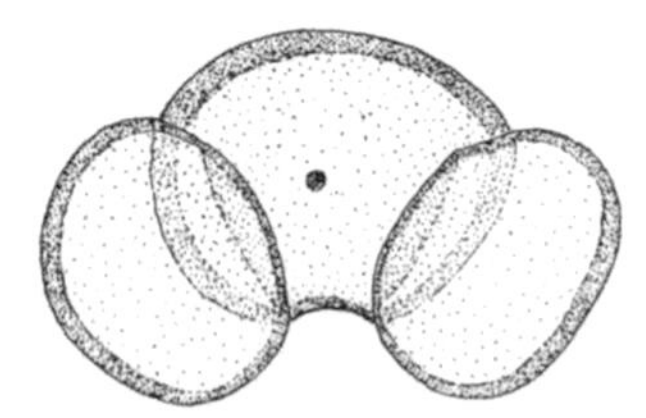

I. 30 Pine pollen grain

I. 31 Copepod spermatophore

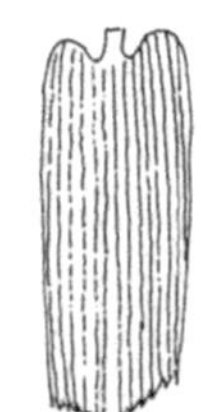

I. 32 Moth scale

I.28–I.32 Some objects commonly found amongst *Sphagnum*

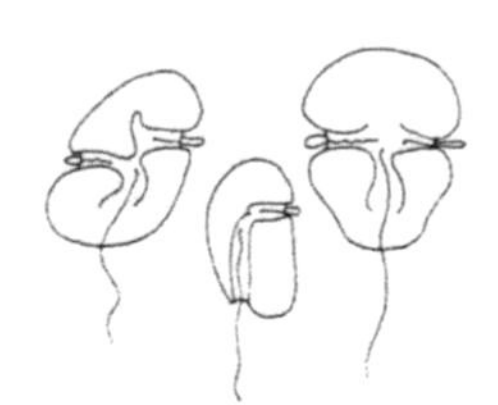

II. 1 Dinoflagellates

Key II. Some flagellates in *Sphagnum*

This key does not cover every possible genus which may be found in the moss. Specimens must conform exactly with the description and figure in order to be identified positively. For further identification see Taylor (1987) for dinoflagellates, Leedale (1967) for euglenoids, and Gojdics (1953) for *Euglena* species.

1 With 2 whip-like hairs or flagella, one of which lies in a groove around the equator of the cell; body grooved and sometimes armoured dinoflagellates (II.1)
– With 1–4 flagella, directed to the front or trailing behind 2

II. 2 *Ochromonas*

2 Coloured 3
– Colourless 15

3 Pigment pale brown or yellowish 4
– Pigment green 7

4 Cells solitary 5
– Cells in group, forming a motile colony 6

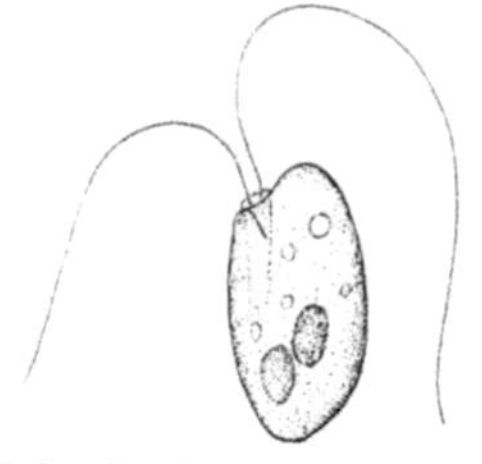

II. 3 *Cryptomonas*

5 Two flagella almost equal in length; body 20 µm or less *Ochromonas* (II.2)
– Two flagella unequal in length; body usually more than 25 µm long (10–80 µm) *Cryptomonas* (II.3)

PLATE 1

Some species of *Sphagnum*

1. *S. magellanicum* Bride
2. *S. capillifolium* (Ehrhart) Hedwig
3. *S. recurvum* Beauvois
4. *S. papillosum* Lindberg
5. *S. cuspidatum* Hoffman

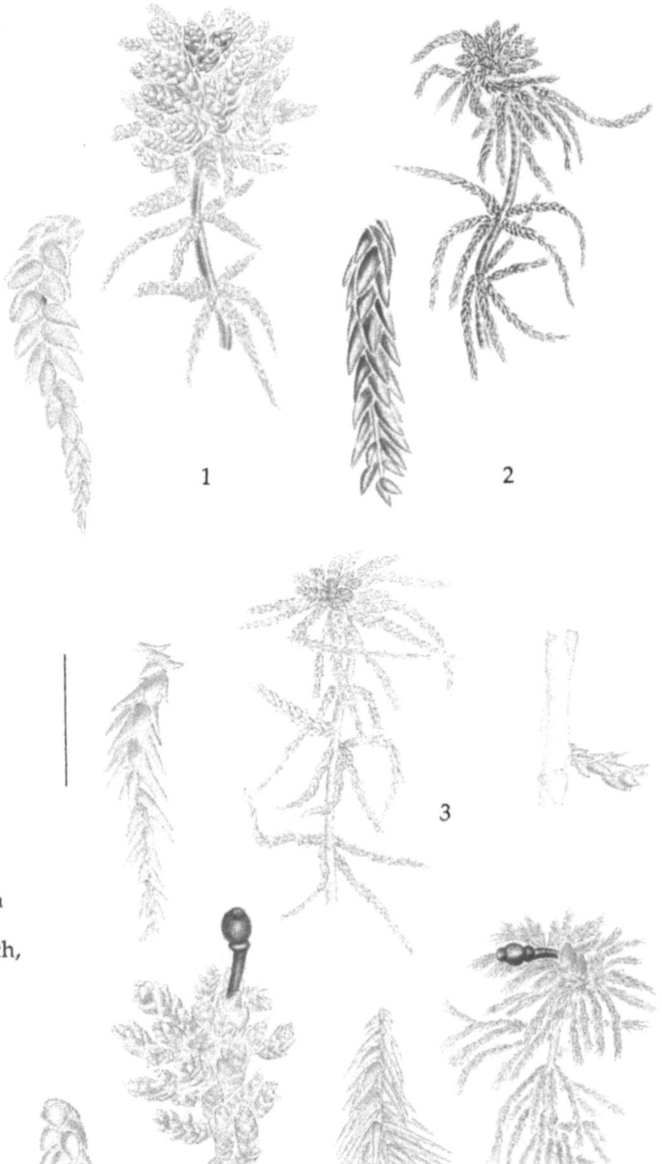

Scale line represents 10 mm
for whole plant
or 5 mm for enlarged branch,
and portion of stem
for *S. recurvum*

PLATE 2

1. Part of leaf of *Sphagnum cuspidatum*, much enlarged

Desmids

2. *Netrium digitus* (Ehrenberg) Itzigs & Rothe
3. Three species of *Closterium*
4. *Micrasterias*
5. *Xanthidium antilopeum* (Brébisson) Kützing

Flagellates

6. *Phacus*
7. *Trachelomonas*
8. *Synura*
   (a) colony x 300
   (b) single cell
   (c) scale of *Synura sphagnicola* Korschikov, x 2500

(6, 7 and 8(a) after Guthrie, 1989; 8(b) and (c) after Pentecost, 1984)

Upper scale line represents 100µm for 2, 3 and 4
Lower scale line represents 100 µm for 5, 6, 7 and 8b

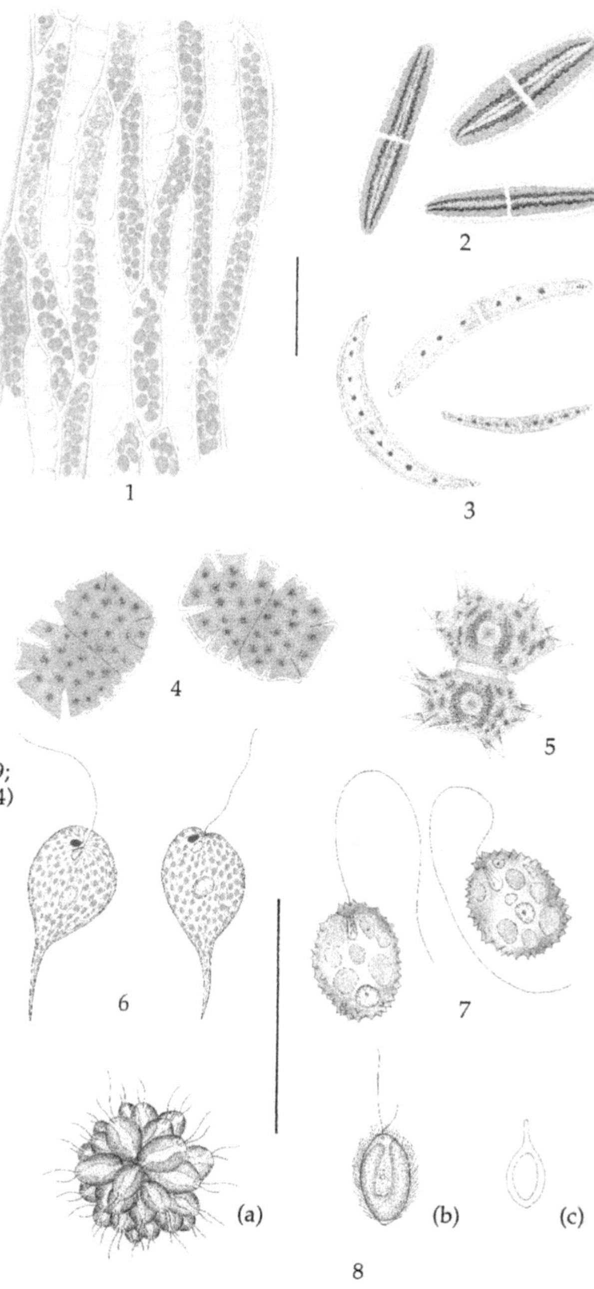

PLATE 3

Testate rhizopods

1. *Bullinularia indica* Penard
2. *Centropyxis aculeata* (Ehrenberg)
3. *Trigonopyxis arcula* (Leidy)
4. *Heleopera rosea* Penard
5. *Hyalosphenia papilio* Leidy
6. *Hyalosphenia elegans* Leidy
7. *Nebela flabellulum* Leidy
8. *Lesquereusia spiralis* (Ehrenberg) Bütschli
9. *Arcella catinus* Penard
10. *Quadrulella symmetrica* (Wallich)

(All from Corbet, 1973)

Scale line 100 μm for 1–8 and 50 μm for 9, 10

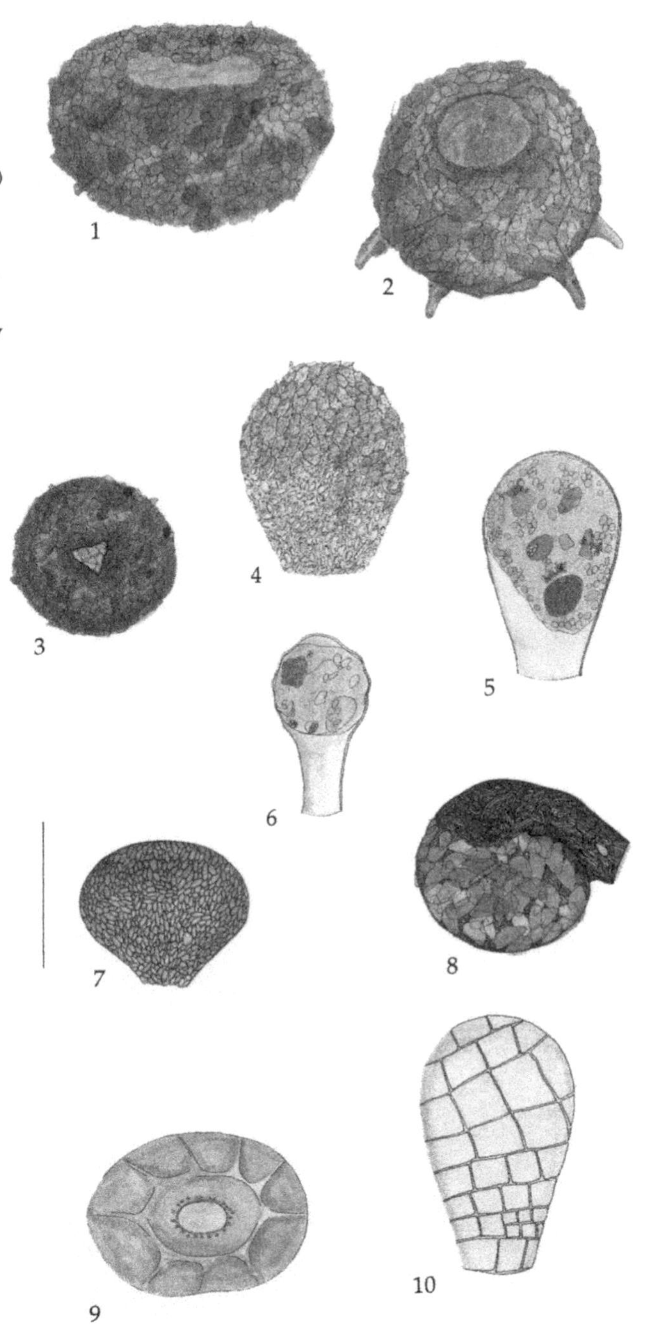

PLATE 4

1. *Arcella discoides* Ehrenberg

2. *Amphitrema wrightianum* Archer

3. *Habrotrocha angusticollis* (Murray) in its case in the axil of a *Sphagnum* leaf

4. Two individuals of *Ophrydium versatile* Müller, a colonial peritrich
   (a) contracted
   (b) expanded

Mites

5. *Hermannia gibba* (Koch)

6. *Ceratoppia bipilis* (Hermann)

7. *Steganacarus magnus* (Nicolet)

8. A galumnid mite

Scale lines 50 µm for 1–4 and 500 µm for 5–8

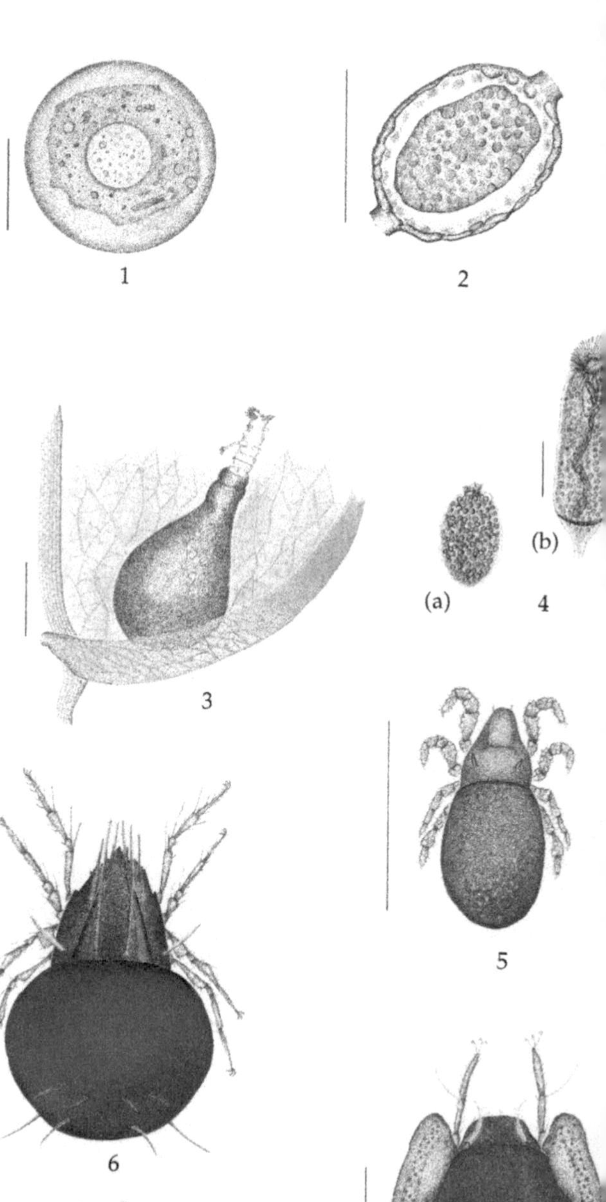

A. Diagram to show life history of *Perone dimorpha* Pascher, epiphytic on green cells of *Sphagnum* leaf (after Pascher, 1932)

1. Fully grown vegetative cell with honey-comb structure
2. Cell contents divide
3. Cell contents form small naked amoebae or, more rarely,
4. Biflagellate swarmers
5. Young vegetative cell
6. Small amoeboid form which can creep over leaf surface
7. Large amoeboid form which catches large prey organisms using blunt pseudopodia and smaller prey using long branched rhizopodia
8. Growing vegetative cells

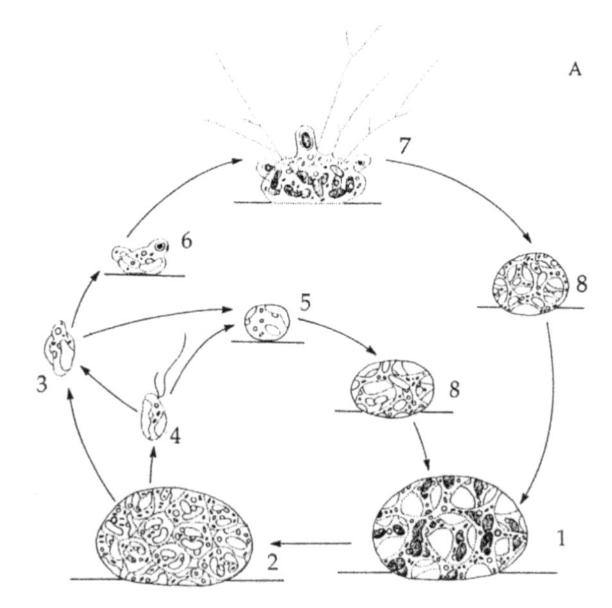

B. Diagram to show life history of *Myxochloris sphagnicola* Pascher (adapted from Pascher, 1930)

1. Flagellate and amoeboid stages have entered hyaline cell of *Sphagnum* leaf by a pore (P)

2-4. These grow and unite (2, 3) to form a large plasmodium (4) which fills the cell

5. (a) A thick-walled cyst develops from the plasmodium or (b) the plasmodium divides into thin-walled spores
6. The cyst may divide into smaller cysts
7. Each spore (a) or cyst (b) sets free flagellate or amoeboid forms
8. These invade another hyaline leaf cell

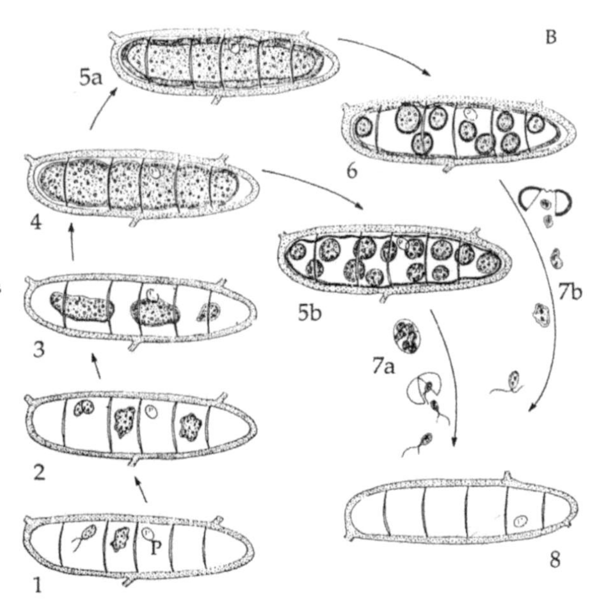

PLATE 6

Ciliate protozoans

1. *Stentor*
2. *Paradileptus*
3. *Dileptus*
4. *Paramecium bursaria*
5. *Oxytricha*
6. *Vorticella*

Colonial flagellates

7. *Platydorina*
8. *Dinobryon*
(after Pentecost, 1984)
(a) single individual enlarged,
(b) colony

Scale lines all 100 μm

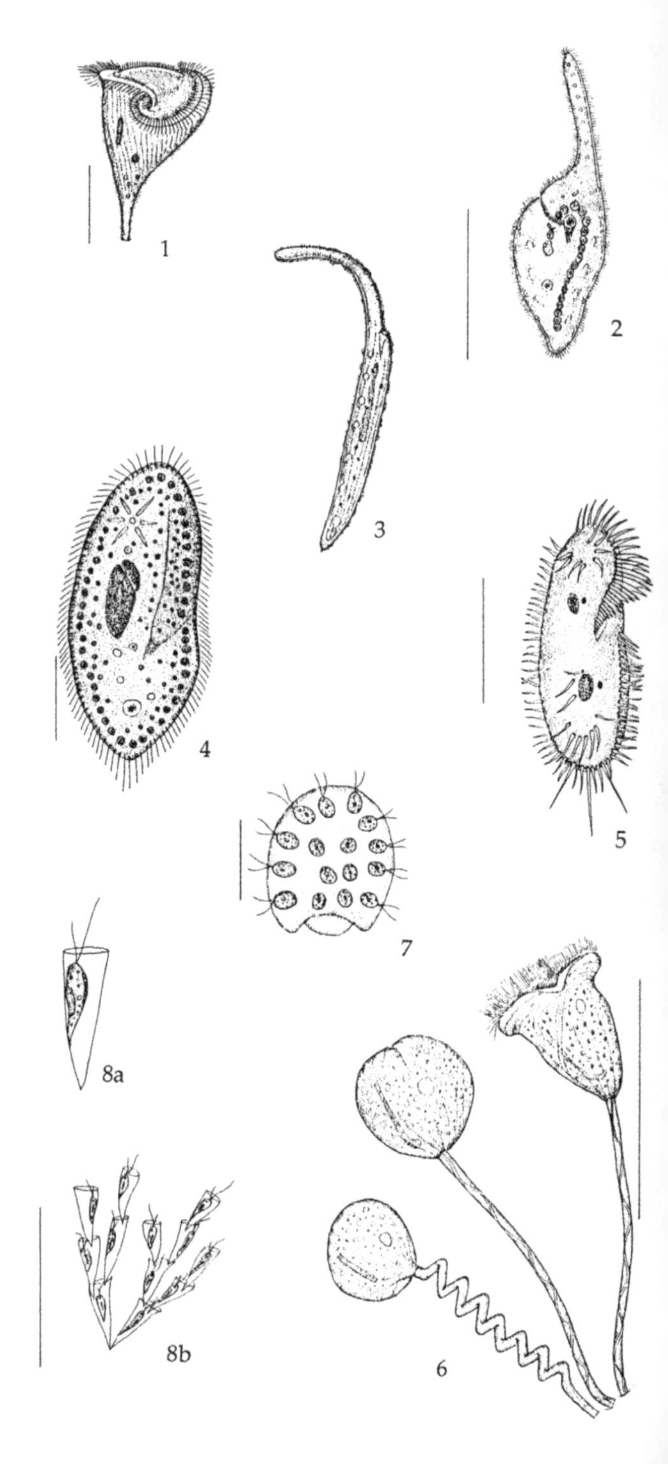

PLATE 7

Ploimate rotifers

1. *Cephalodella gibba* (Ehrenberg)
2. *Colurella adriatica* Ehrenberg
3. *Elosa woralli* Lord (after Voigt, 1956) (a) side view showing distinctive crescentric opening (lower left), (b) front view showing eye spot and trophi (jaws) of unequal length
4. *Polyarthra minor* Voigt
5. *Lepadella ovalis* (Müller)
6. Lorica of *Keratella serrulata* (Ehrenberg)
7. *Lecane*

Scale lines all 100 μm

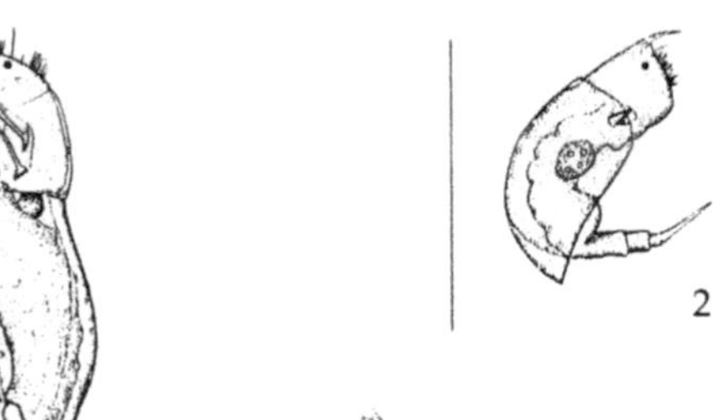

2

1

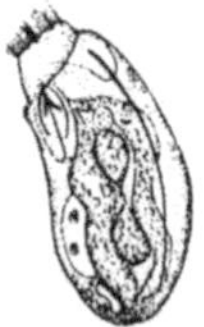

3a

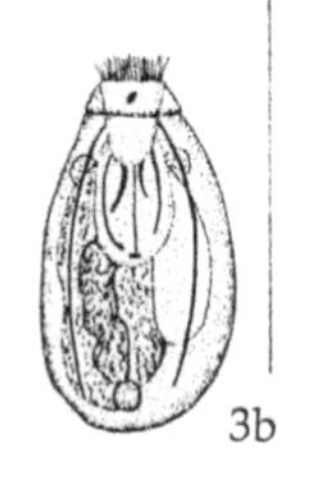

3b

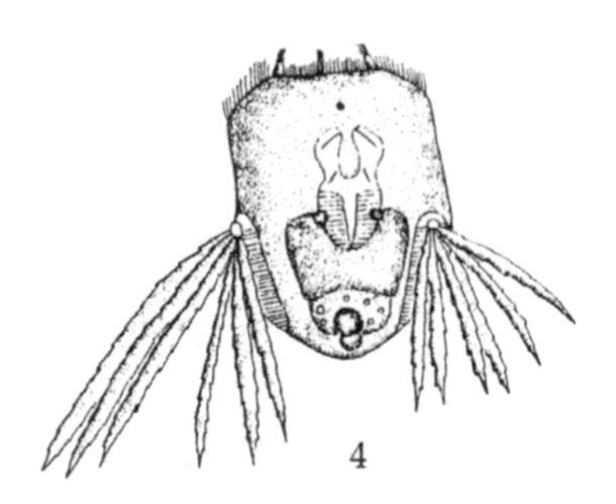

4

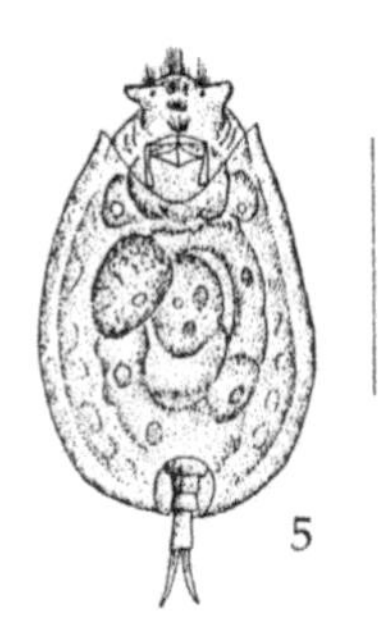

5

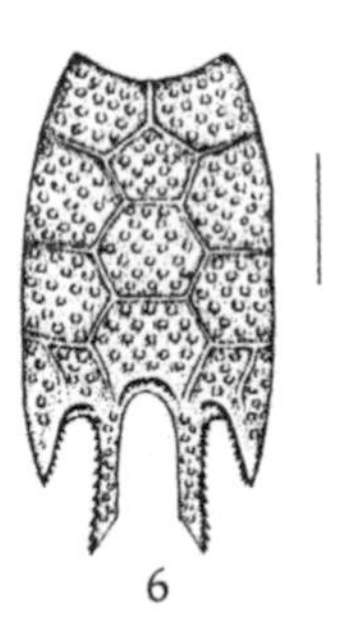

6

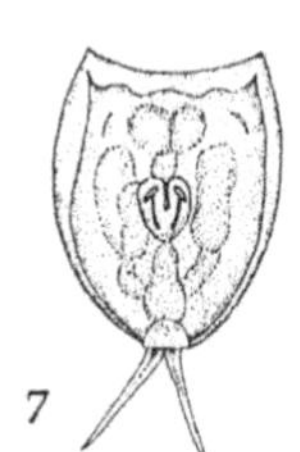

7

PLATE 8

Mites

1. The water mite *Hydrovolzia placophora* (Monti) (after Gledhill, 1960)
   (a) from above
   (b) from below

2. The mesostigmatid mite *Cilliba cassidea* (Hermann)

Cladoceran crustaceans

3. *Acantholeberis curvirostris* (Müller)

4. *Simocephalus vetulus* (Müller)

5. *Acroperus harpae* Baird

6. *Chydorus sphaericus* (Müller)

7. *Graptoleberis testudinaria* (Fischer)

(3–7 after Ward & Whipple, 1918; antennae not shown in 3, 5 and 7)

Scale lines all 500 µm

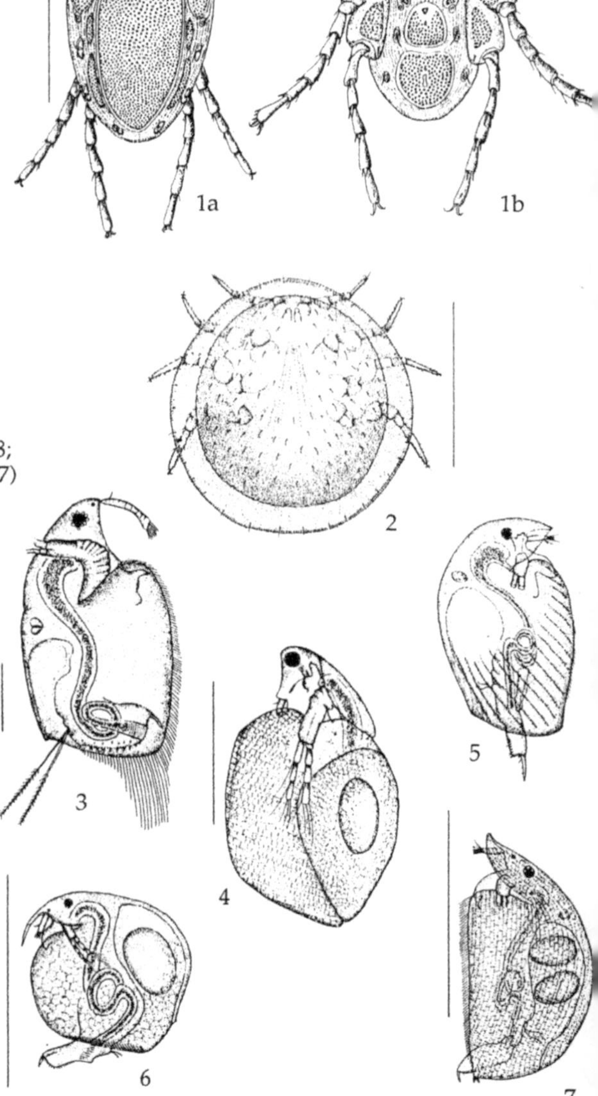

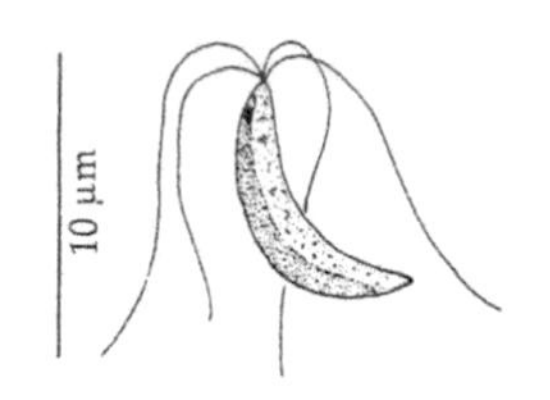

II. 4 *Spermatozopsis*

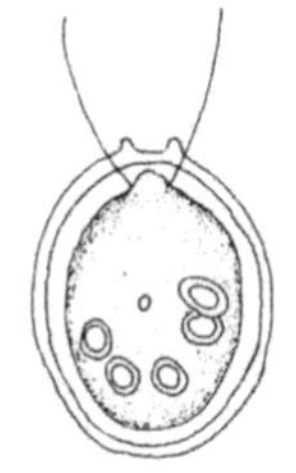

II. 5 *Chlamydomonas sphagnicola*

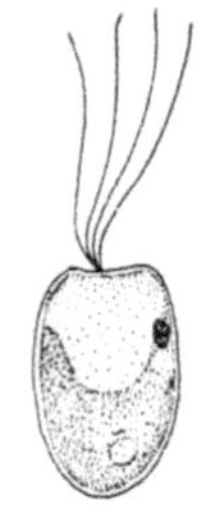

II. 6 *Carteria*

II. 7 *Lepocinclis* (10–30 µm long)

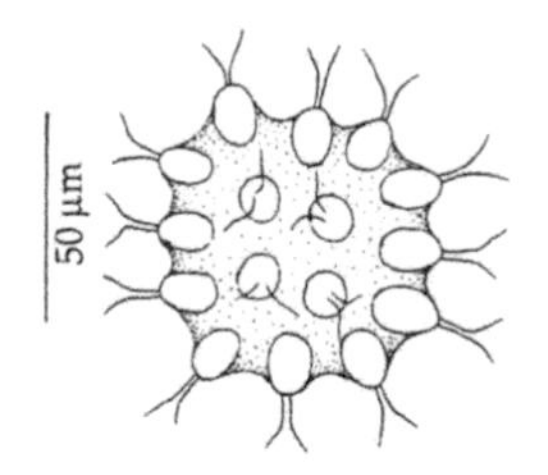

II. 8 *Gonium pectorale*

6 Each cell of a colony enclosed within a delicate case; 2 flagella of unequal length — *Dinobryon* (pl. 6.8)
– Cells not so enclosed; 2 flagella of equal length — *Synura* (pl. 2.8)

7 Cells solitary — 8
– Cells in a group forming a motile colony — 14

8 Body enclosed in brown pot-like container; one flagellum — *Trachelomonas* (pl. 2.7)
– Body not enclosed in this way — 9

9 Body sickle-shaped; 4 flagella; very small, about 8 µm — *Spermatozopsis* (II.4)
– Body not this shape — 10

10 With 2 or 4 flagella; body oval or spherical — 11
– With 1 flagellum; body spherical or elongate — 12

11 With 2 flagella — *Chlamydomonas* (II.5)
– With 4 flagella; 4–70 µm long — *Carteria* (II.6)

12 Strongly flattened; body broad next to flagellum; tapered at opposite end; 30–170 µm — *Phacus* (pl. 2.6)
– Not shaped like this — 13

13 Body oval; short spine at opposite end to flagellum; 10–30 µm long — *Lepocinclis* (II.7)
– Much elongated; 25–500 µm long; many species, some very common amongst *Sphagnum* — *Euglena*

(Fig. 26 shows a species that lacks a flagellum.)

14 Colony a flat plate of 4–32 cells; flagella emerge from edges and one face only — *Gonium* (II.8)
– Colony a flat plate of 16–32 cells; flagella emerge from edges and both faces — *Platydorina* (pl. 6.7)

Other colonial forms may also occur.

15 With finger-like pseudopodia and also 1 flagellum — *Mastigella* or *Mastigamoeba* (II.9)
– Without pseudopodia; with 1 or 2 flagella — 16

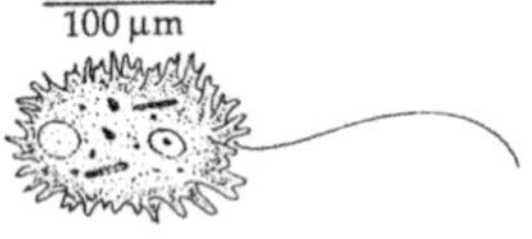

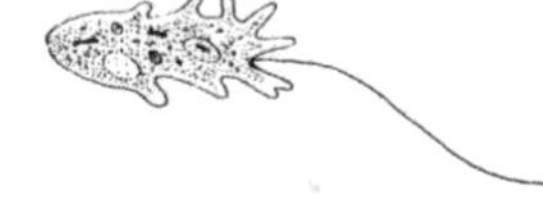

II. 9 (a) *Mastigella* (b) *Mastigamoeba* (20–200 µm long)

16 With 1 flagellum visible; body changeable in form, often spindle-shaped and pointed behind; 20–70 µm long *Astasia* (II.10)

– With 2 flagella 17

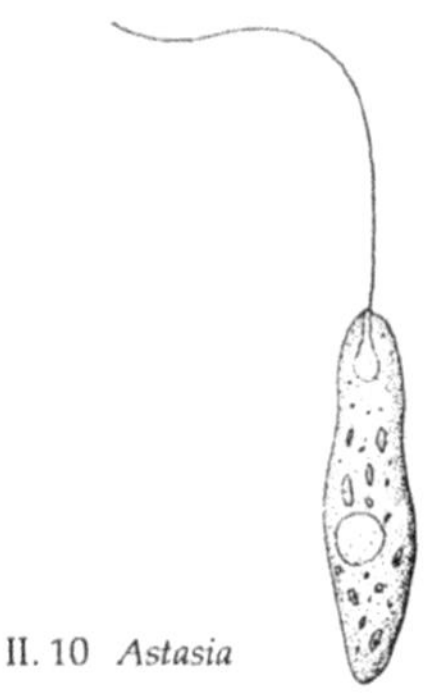

II. 10 *Astasia*

17 With 2 flagella of unequal length, both directed forwards. Body undergoes rapid change in shape, showing an extreme form of euglenoid movement *Distigma proteus* (II.11)

– With 2 flagella; one directed forwards and the other trailing 18

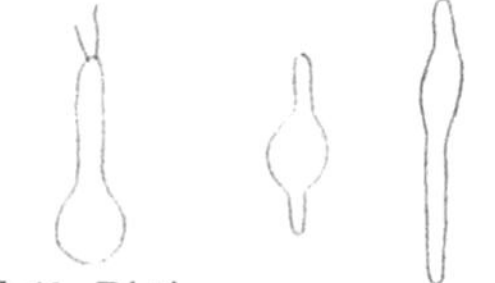

II. 11 *Distigma*: euglenoid movement

18 With light-scattering bodies; body about 50 µm long *Gonystomum* (II.12)

Without light-scattering bodies 19

19 Body usually over 25 µm long; hind end broad; trailing flagellum may be hidden; very common; 20–70 µm long *Peranema* (II.13)

– Body 25 µm long or less (if in doubt check II.10, II.13, II.14 and II.15) 20

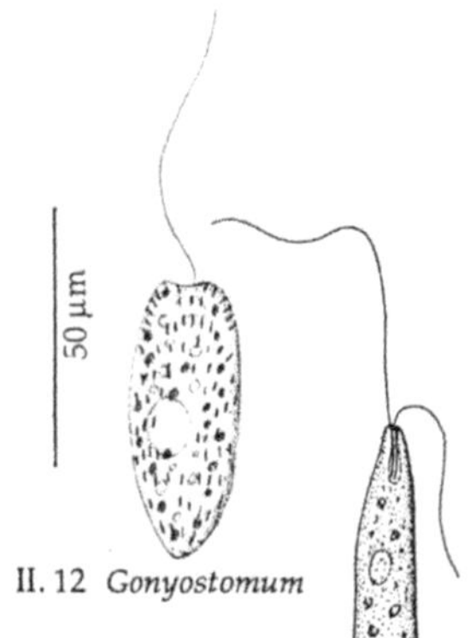

II. 12 *Gonyostomum*

II. 13 *Peranema*

20 With long slit-like furrow; trailing flagellum short; body 8–25 µm long *Notoselenus* (II.14)

– Without a long furrow 21

21 Oval and flattened; many short furrows at front end; body about 25 µm long *Entosiphon* (II.15)

– Differently shaped and very small (less than 20 µm long) 22

22 Usually attached to substrate by trailing flagellum; body about 8 µm long *Pleuromonas* (II.16)

– Not attached; body about 15 µm long and capable of changing shape *Bodo* (II.17)

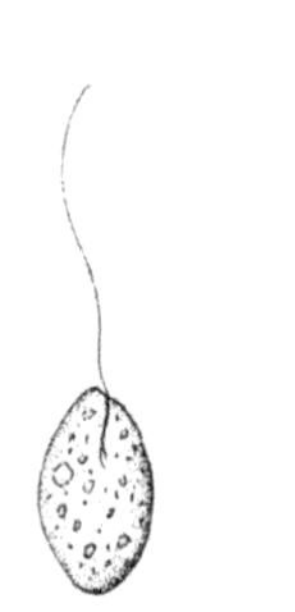

II.14 *Notosolenus*

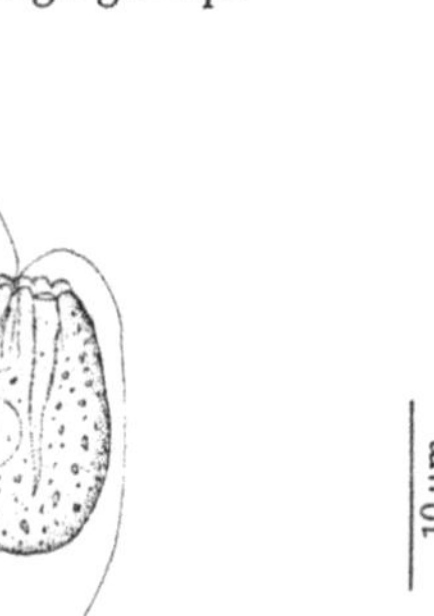

II. 15 *Entosiphon*

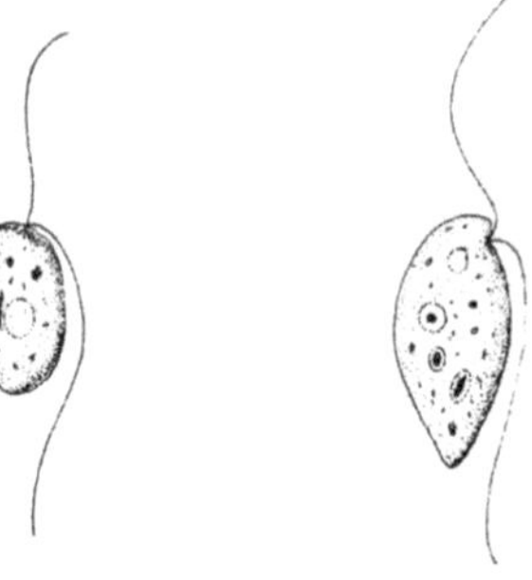

II. 16 *Pleuromonas*

II. 17 *Bodo*

Key III. Some genera of desmids in *Sphagnum*

This key does not include all genera found in Sphagnum. Specimens must conform precisely to the description and figure for positive identification. For a more detailed key see Lind & Brook (1980).

1	Cells joined to form filaments	2
–	Cells solitary or in non-filamentous groups	4
2	Cells slightly constricted but without a distinct notch across the middle	*Hyalotheca* (III.1)
–	Cells with a distinct notch (isthmus) across the middle (see III.10)	3
3	Isthmus shallow; cells joined by broad, flat tapering ends	*Bambusina* (III.2)
–	Cells with deep isthmus and joined by mucilaginous pads which may or may not be visible	*Spondylosium* (III.3)
4	Cells without a notch (isthmus) across the middle	5
–	Cells with a slight or distinct isthmus	8
5	Cells short, cylindrical, 1.5–2.5 times longer than broad	6
–	Cells more elongated; cylindrical, wider in the middle and tapering at each end (fusiform), rod-shaped or crescent-moon shaped (lunate)	7
6	With single, simple ribbon-like chloroplast	*Mesotaenium* (III.4)
–	With 2 star-like or ridged chloroplasts, one at each end of the cell	*Cylindrocystis* (III.5)
7	Cells rod-shaped, or more commonly lunate with tapering ends; one ridged chloroplast in each cell half	*Closterium* (III.6, pl. 2.3)
–	Cell fusiform, tapering to blunt ends; a complex chloroplast in each half cell, with 6 radiating longitudinal plates	*Netrium* (pl. 2.2)
8	Cells with a slight isthmus across the middle	9
–	Cells with a very distinct isthmus	11
9	Cells straight, fusiform (wider in centre and tapering at each end), 4–6 times longer than broad and with a distinct cleft at the tip	*Tetmemorus* (III.7)
–	Cells straight, cylindrical and without a cleft at the tip	10

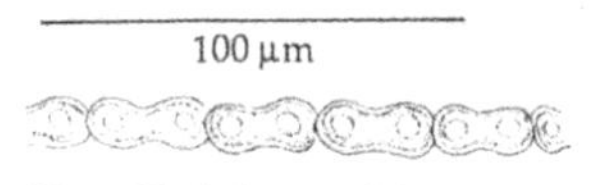

III. 1 *Hyalotheca undulata*

50 µm

III. 2 *Bambusina brebissonii*

from the side

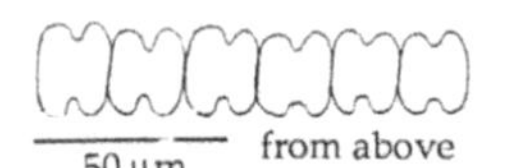

III. 3 *Spondylosium planum* (after Lind & Brook, 1980)

50 µm

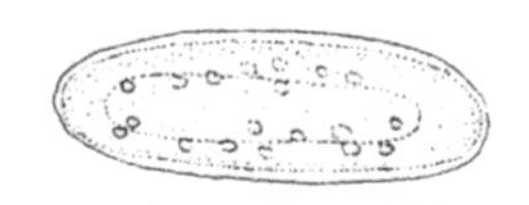

III. 4 *Mesotaenium endlichearum*

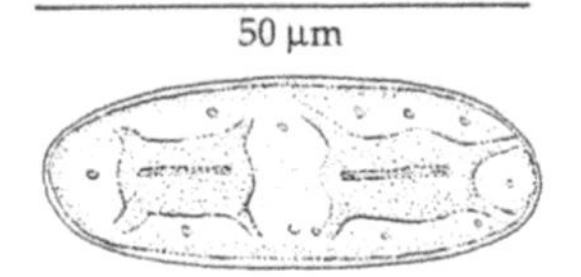

III. 5 *Cylindrocystis brebissonii*

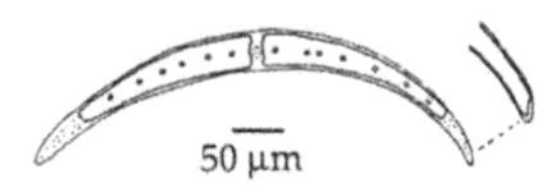

III. 6 *Closterium dianae* (after Lind & Brook, 1980)

50 µm

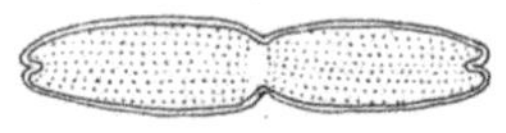

III. 7 *Tetmemorus brebissonii* (after Lind & Brook, 1980)

10 Isthmus shallow; cell walls with sparse pores in regular rows — *Penium* (III.8)
– Isthmus more conspicuous; pores dense in irregular or oblique rows — *Actinotaenium* (III.9)

11 Cells flattened, elliptical when seen end-on — 12
– Cells not flattened, triangular when seen end-on with corners extended into lobes or points — 16

12 Cell outline usually circular or elliptical, sometimes wavy; margins unbroken (except at isthmus); cell wall with pores, smooth or ornamented with warts or granules — *Cosmarium* (III.10)
– Cell margins angular or bearing incisions (cuts), lobes or spines — 13

13 Cells circular to 6-sided in outline with margins always deeply cut and lobed — *Micrasterias* (pl. 2.4)
– Cells differently shaped; either longer than broad or with angular margins — 14

14 Cells mostly distinctly longer than broad; deep isthmus and with a cleft in the middle of the tip of each semi-cell — *Euastrum* (III.11, III.12)
– Cells not longer than broad; cell margins angular — 15

15 Cell margins angular and ornamented by pairs of simple spines; face of cell with central swelling, often with small rounded depressions — *Xanthidium* (III.13, III.14, pl. 2.5)
– Cell angles extend as lobes which bear single spines; walls smooth and unornamented — *Staurodesmus* (III.15)

16 Walls and lobes ornamented with granules, spines or warts — *Staurastrum* (III.16)
– Walls smooth, lobes each tipped with a single spine — *Staurodesmus* (III.15)

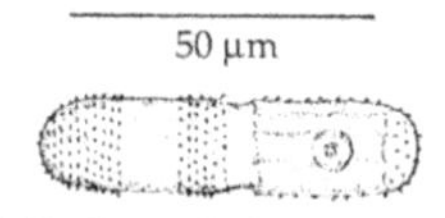

III. 8 *Penium cylindrus* (after Lind & Brook, 1980)

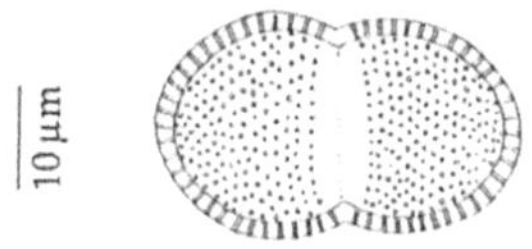

III. 9 *Actinotaenium cucurbitinum* (after Brook, 1990)

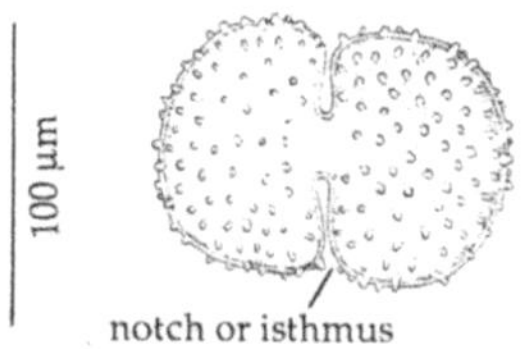

III. 10 *Cosmarium brebissonii*

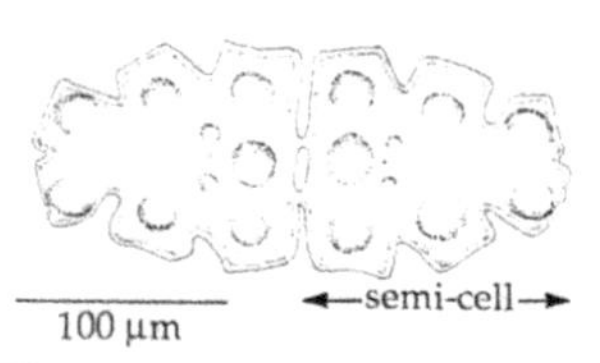

III. 11 *Euastrum oblongum*

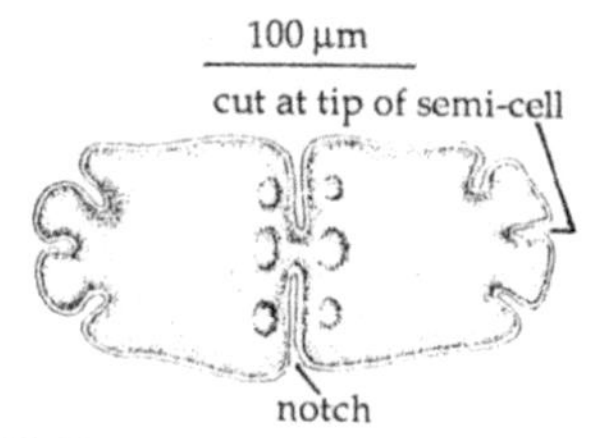

III. 12 *Euastrum crassum*

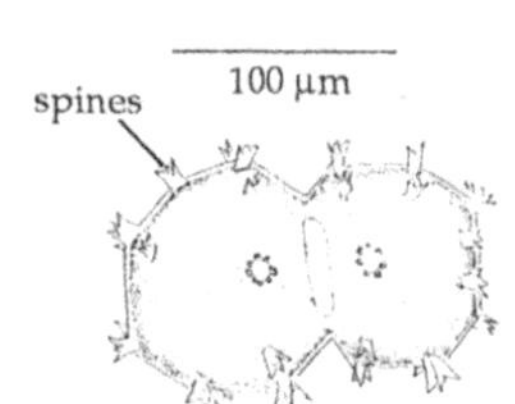

III. 13 *Xanthidium armatum*

100 µm

III. 14 *Xanthidium antilopeum*, just after cell division

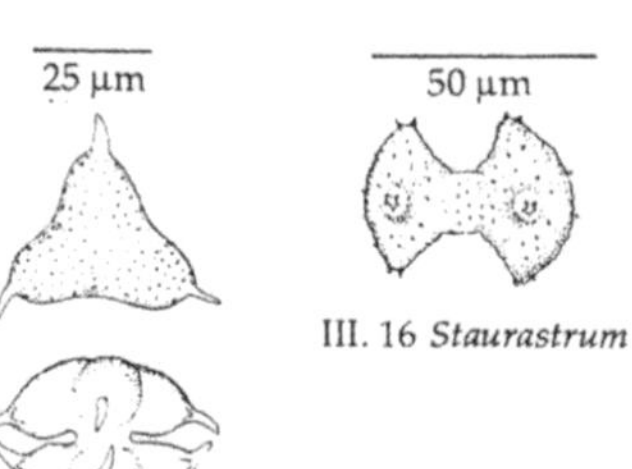

III. 15 *Staurodesmus dickei* (after Lind & Brook, 1980)

III. 16 *Staurastrum*

Guide I. Some genera and species of diatoms in *Sphagnum*

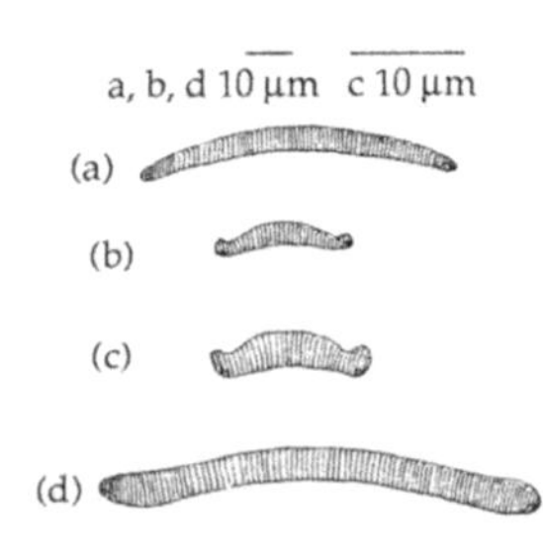

GI. 1 (a) *Eunotia curvata*
(b) *E. tenella*
(c) *E. exigua*
(d) *E. parallela*

Although the bog diatom flora is well known and extensive, the species most closely associated with *Sphagnum* throughout Britain are not well recorded. The identification of diatoms is difficult because it depends so much on the detailed ornamentation of the siliceous outer casing, or frustule (see p. 10). To see this properly requires techniques beyond the scope of a non-specialist: careful cleaning of the diatoms (Barber & Haworth, 1981), together with high power phase contrast microscopy. Frequent changes in the classification and naming of genera and species compound the problem. Some of the commoner or more distinctive genera or species are described here, using features which can be seen without cleaning the diatoms or by simply clearing them in Hoyer's medium (p. 55). See Barber & Haworth (1981) for detailed description of the diatom frustule, Krammer & Lange-Bertalot (1986–) for descriptions and drawings of species, and Round, Crawford & Mann (1990) for more recent information.

Eunotia is the genus best represented and most easily recognisable. There are often many individuals of several species in a squeeze of the moss (GI.1). The valves are not bilaterally symmetrical; they are often curved and have a distinct mark or node near each end. The ends are sometimes knobbed.

The valves of *Pinnularia* (GI.2, GI.3) are often large (up to 200 µm in some species) and cigar-shaped, sometimes with knobbed ends. The central line or raphe is often wide and strap-like. The transverse markings are broad and finger-like with no internal patterns.

The genus *Navicula* (GI.4) is sometimes difficult to distinguish from *Pinnularia*. Most of the species occurring in *Sphagnum* are small (10–30 µm). The central raphe is narrow and thread-like. The markings are fine or thick and usually composed of rows of dots. Some species of *Navicula* have now been placed in other genera; *N. variostriata* Krasske is now *Calvinula variostriata* (Krasske) Mann.

Frustulia rhomboides is a typical species of the bog flora (GI.5) and common amongst *Sphagnum*. The central raphe lies within a rib-like thickening. The fine parallel transverse markings are crossed by fine longitudinal lines.

Brachysira serians (GI.6) can be distinguished by the fine transverse patterning which is distinctly flecked.

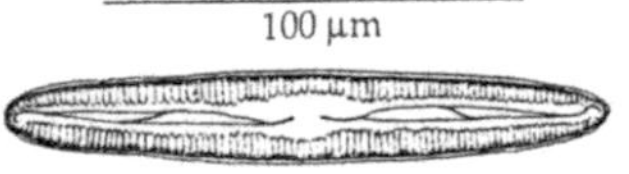

GI. 2 *Pinnularia viridis* (Nitzsch) Ehrenberg

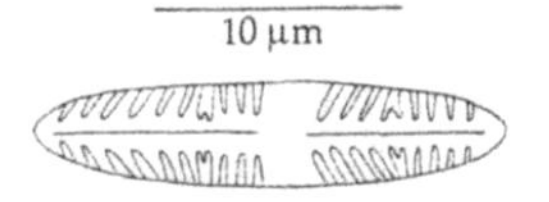

GI. 3 *Pinnularia acoricola* Hustedt (after Denys)

10 µm

GI. 4 *Navicula subtilissima* Cleve

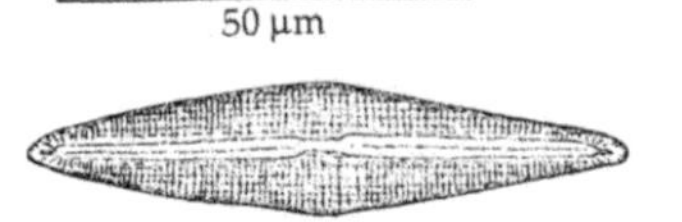

GI. 5 *Frustulia rhomboides* (Ehrenberg) de Toni

GI. 6 *Brachysira serians* (Kützing) Round & Mann

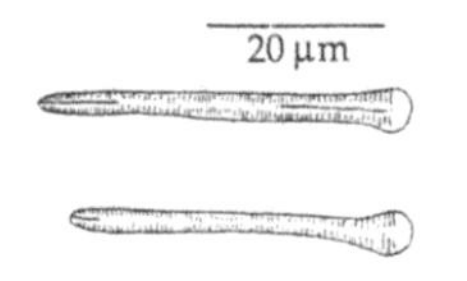

GI. 7 *Peronia fibula* (Kützing) Ross (after Barber & Haworth, 1981)

Peronia fibula (GI.7) (previously known as *P. heribaudi*) is wedge-shaped and grows on the leaves of *Sphagnum*.

Surirella biseriata is a large and very distinctive species (GI.8) occasionally found in *Sphagnum* but not typical of the habitat.

A few diatom species are colonial; *Tabellaria flocculosa* (GI.9) is common in *Sphagnum* and has long, often zigzag chains. *Fragilariforma virescens* (previously known as *Fragilaria virescens* Ralfs) (GI.10), with its ribbon-like chains, is less commonly found.

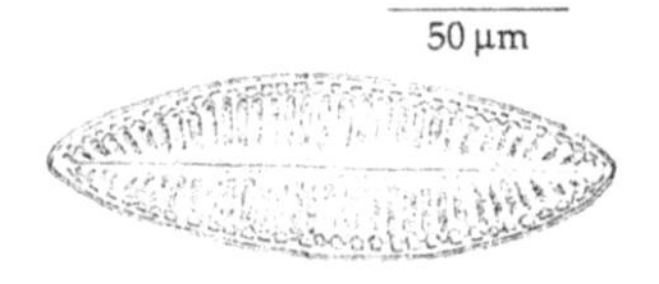

GI. 8 *Surirella biseriata* Brébisson

GI. 9 *Tabellaria flocculosa* (Roth) Kützing, valve view (left) and two cells in girdle view (right)

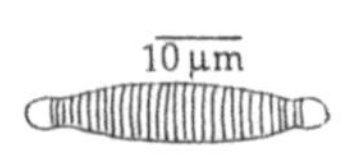

GI. 10 *Fragilariforma virescens* (Ralfs) Williams & Round, single frustule of chain

Key IV. Some genera of testate rhizopods in *Sphagnum*

For further details and help with identification refer to Corbet (1973), from which this key is adapted, Cash, Wailes & Hopkinson (1905–21), or Grospietsch (1965).
This key does not include all the genera found in *Sphagnum*. Specimens must conform precisely to the descriptions and illustrations for positive identification.

1	Test of secretion without visible plates or particles, teeth or spines	2
–	Test with plates or particles	7
2	Test a domed disc or hemisphere with a central mouth (test usually yellow-brown with a finely-patterned surface)	*Arcella* (IV.1, pl. 4.1)
–	Test of a different shape; mouth not central	3
3	Test oblong with a small mouth at each end	*Amphitrema flavum* (Archer) (1V.2)
–	Only one mouth	4

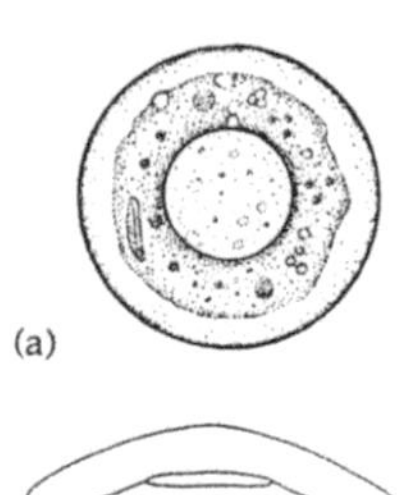

IV. 1 (a) *Arcella discoides*, living animal inside test; (b)–(d) side views of 3 species, 100–150 µm diameter
(b) *A. discoides*
(c) *A. catinus*
(d) *A. vulgaris*
(after Deflandre, 1928)

green zoochlorellae
mouth
mouth
amber-coloured transparent test
25 µm

IV. 2 *Amphitrema flavum*

4 Test brown or amber and less than 30 µm long *Cryptodifflugia* (IV.3)

– Some other colour, or if brown more than 30 µm long 5

5 Mouth at end of test, at right angles to long axis *Hyalosphenia* (IV.4, IV.5)

– Mouth oblique or not quite at end of test 6

6 Test up to 50 µm long *Trinema* or *Corythion* (IV.6, IV.7)

The genera *Trinema* and *Corythion* include very small species with tests so transparent that their plates are often difficult or impossible to see unless a phase contrast microscope is available. To separate the two genera it is necessary to see whether the plates touch one another (*Trinema*) or not (*Corythion*)

– Test more than 60 µm long *Cyphoderia* (IV.8)

7 Test covered with foreign particles 8

– Test without untidily stuck-on particles 14

8 Test appearing asymmetrical, with a curved tube on one side leading to the mouth *Lesquereusia* (pl. 3.8)

– Not like this 9

9 Test oblong with a small mouth at each end *Amphitrema stenostoma* Neusslin or *A. wrightianum* Archer (IV.9, pl. 4.2)

– One mouth only 10

10 Test neat, the hindmost part decorated with larger particles *Heleopera* (pl. 3.4)

– Particles similar throughout test 11

11 Mouth at the end of test *Difflugia* (IV.10)

– Mouth at one side of test 12

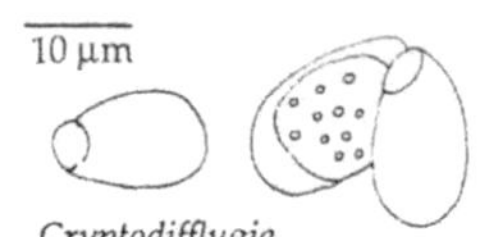

IV. 3 *Cryptodifflugia*

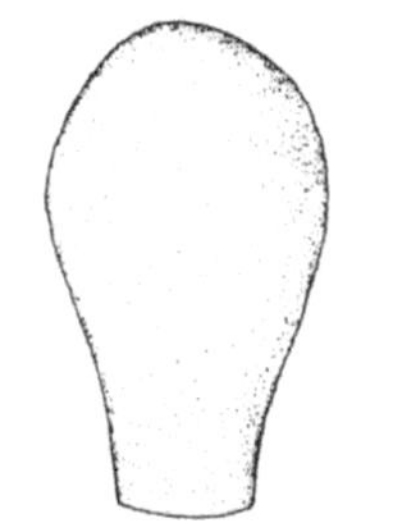

IV. 4 *Hyalosphenia papilio*

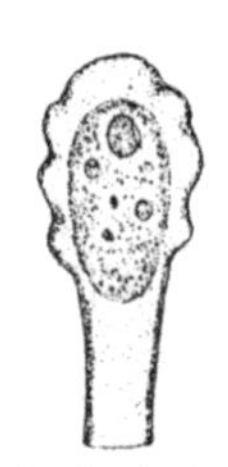

IV. 5 *Hyalosphenia elegans*

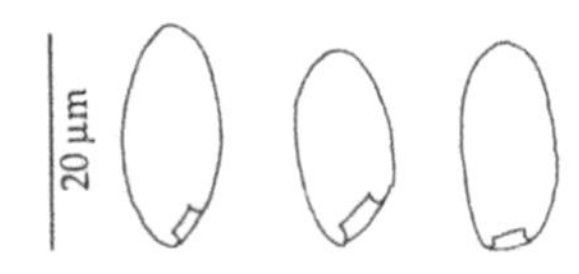

IV. 6 *Trinema lineare* (after Corbet, 1973)

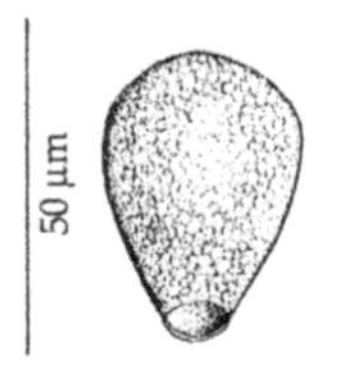

IV. 7 *Corythion dubium* (after Corbet, 1973)

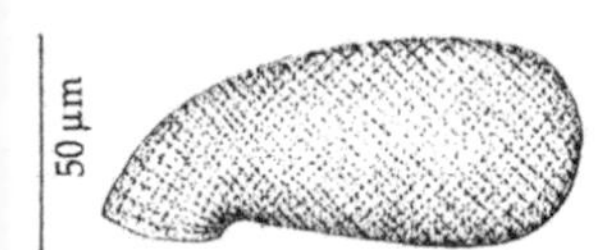

IV. 8 *Cyphoderia ampulla* (after Corbet, 1973)

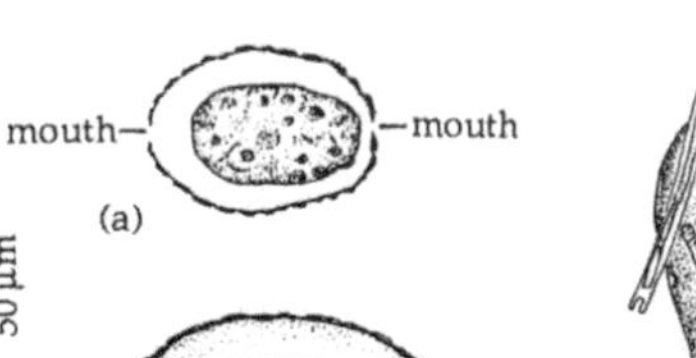

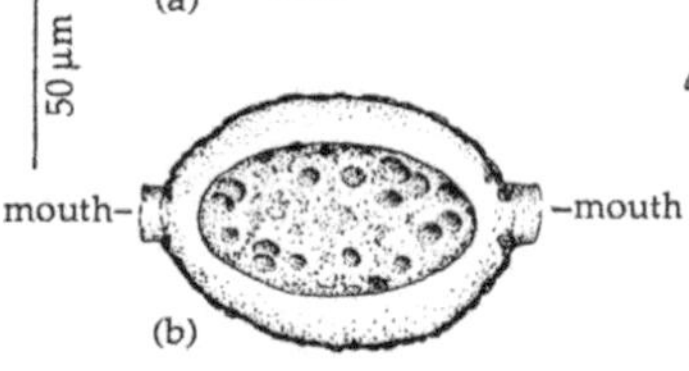

IV. 9 (a) *Amphitrema stenostoma* (b) *A. wrightianum*

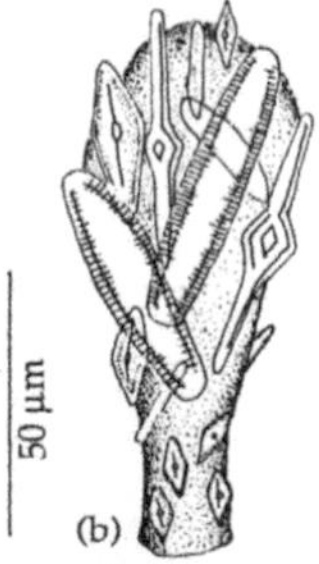

IV. 10 (a) *Difflugia rubescens*; (b) *D. bacillifera* (after Cash, Wailes & Hopkinson, 1905–21)

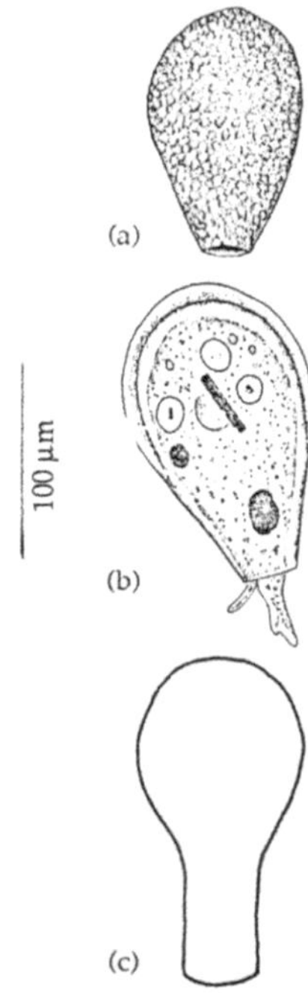

IV. 11 (a) *Nebela tincta*
(b) *N. carinata*
(c) outline of *N. lageniformis*

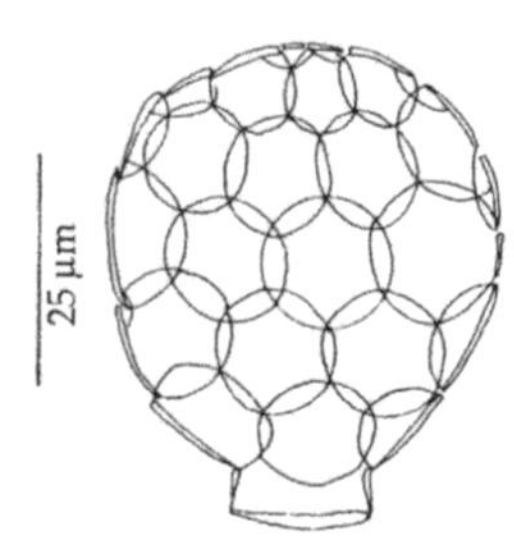

IV. 12 *Sphenoderia lenta*

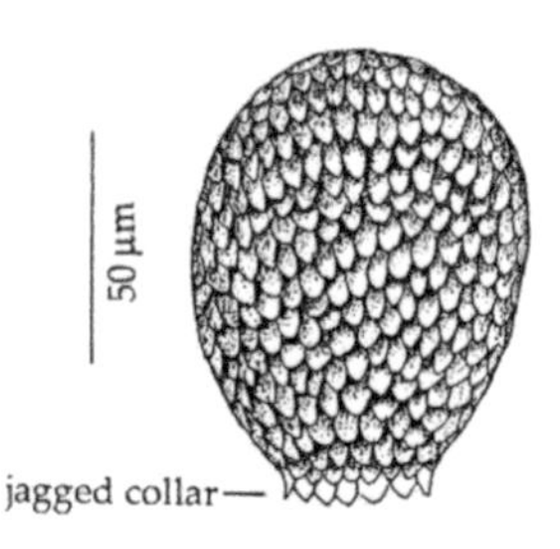

IV. 13 *Assulina muscorum*

12 Mouth a curved slit with an irregular row of pores beside it — *Bullinularia* (pl. 3.1)
– Mouth not slit-shaped; no pores — 13

13 Mouth 3- or 4-sided — *Trigonopyxis* (pl. 3.3)
– Mouth oval, round or crescent-shaped — *Centropyxis* (pl. 3.2)

14 Mouth oblique or not quite at end of test — 15
– Mouth symmetrically placed at end of test — 16

15 Test less than 50 µm long (see couplet 6) — *Corythion* or *Trinema*
– Test more than 60 µm long (see couplet 6) — *Cyphoderia*

16 Plates 4-sided in a neat mosaic — *Quadrulella* (pl. 3.10)
– Plates not 4-sided — 17

17 Plates in a neat mosaic, but not geometrically patterned; no membranous collar round mouth — *Nebela* (IV.11)
– Without this combination of characters — 18

18 Test with membranous collar (which may be jagged) around mouth; no spines or teeth — 19
– Test without membranous collar; test colourless, flat, sometimes bearing spines — 20

19 Test ovoid or spherical; colourless — *Sphenoderia* (IV.12)
– Test flattened, colourless or brown — *Assulina* (IV.13)

20 Mouth bordered by plates with toothed margins — *Euglypha* (IV.14)
– Mouth plates not toothed — *Placocista* (IV.15)

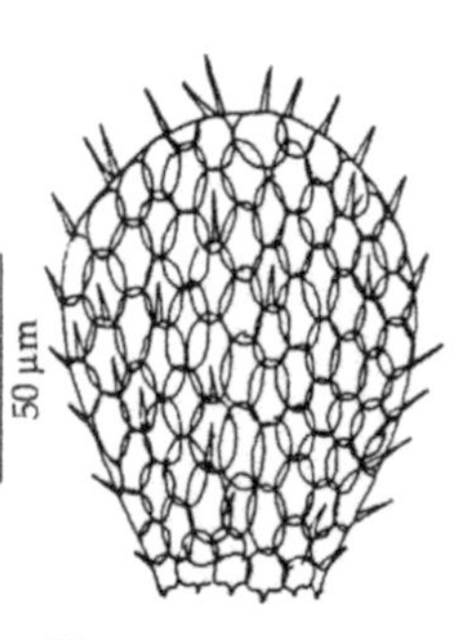

IV. 14 *Euglypha ciliata*

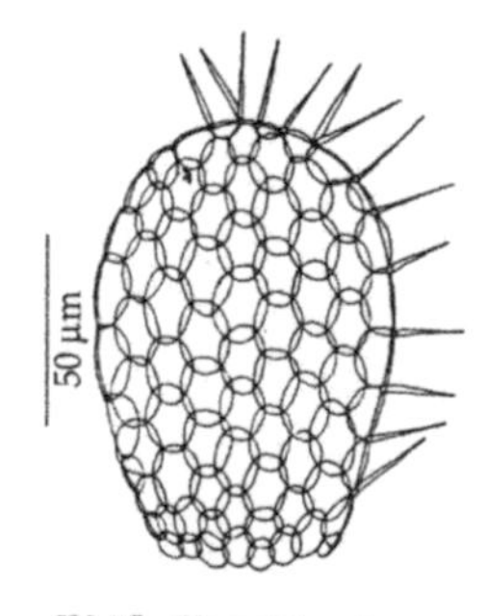

IV. 15 *Placocista spinosa* (after Leidy, 1879)

Key V. Rotifers in *Sphagnum*: main groups and some genera and species

Thorp & Covich (1991) review the ecology of rotifers and provide a key to families.

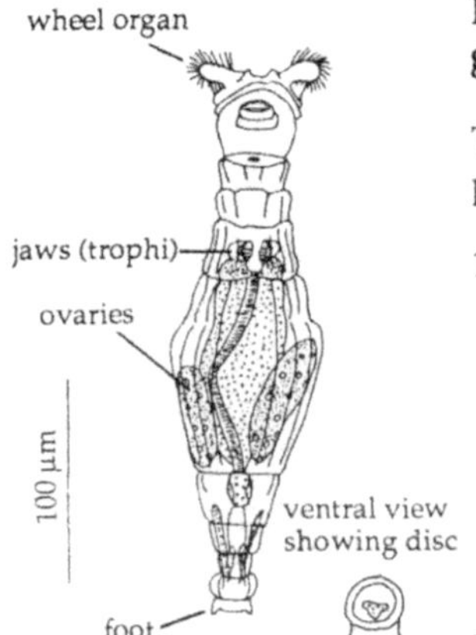

V. 1 *Mniobia*

1 Body cylindrical, highly contractile and telescopic; without a stiffened skin or lorica. Free-swimming or crawling in a leech-like manner. Foot with 2 spurs and either a sucker-like disc or 2 to 4 toes, seen only when body is fully extended. Ovaries paired; wheel organ sometimes appears as 2 separate wheels; some species very common in *Sphagnum* Order Bdelloidea (V.1) 2

Bartos (1951) gives descriptions and drawings of European bdelloids, including many from *Sphagnum* in Britain.

– Body without this combination of characters Order Monogononta

For further identification of the Monogononta see Hudson & Gosse (1889), Donner (1966), Ruttner-Kolisko (1974), Koste (1978) or Pontin (1978). Most of the species found in *Sphagnum* are in the large sub-order Ploima, all of which are free-swimming. Examples are shown in Plate 7 and V.2. Some species from the other two sub-orders live fixed to *Sphagnum* or other plants (V.3).

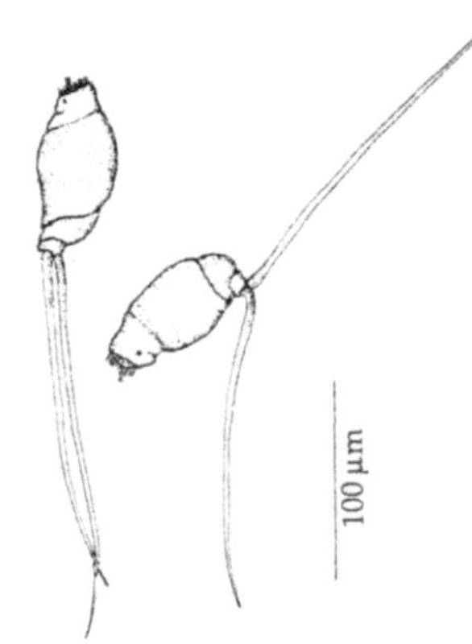

V. 2 *Notommata*, a free-swimming ploimate rotifer

2 Stomach not in the form of a bag or tube. Food and excreta in the form of pellets. Some individuals may carry an egg 3

– Stomach a tube-shaped bag. Food and excreta not in the form of pellets 7

3 Head with well-developed hood (V.4) *Scepanotrocha*

– Head without hood *Habrotrocha* 4

4 Animal living inside *Sphagnum* cells or in a bottle-shaped case between the leaves 5

– Free-living forms several species

5 Animal living in a bottle-shaped case between the leaves *H. angusticollis* (Murray) (V.5, pl. 4.3)

– Animal living in *Sphagnum* retort cell of cortex 6

Sometimes other species of bdelloid rotifers are found in hyaline leaf cells.

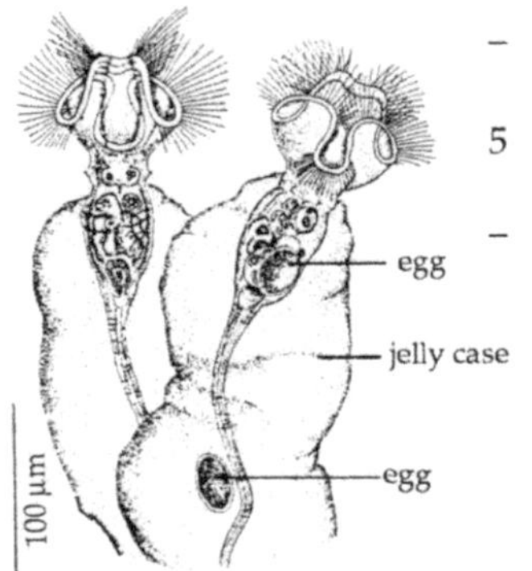

V. 3 *Collotheca trilobata*, a rotifer living on *Sphagnum* and other plants (after Hudson & Gosse, 1889)

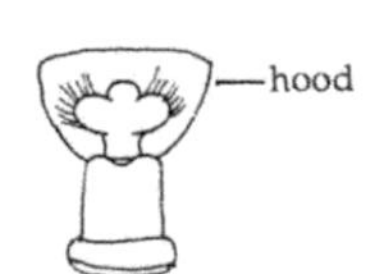

V. 4 Head of *Scepanotrocha*

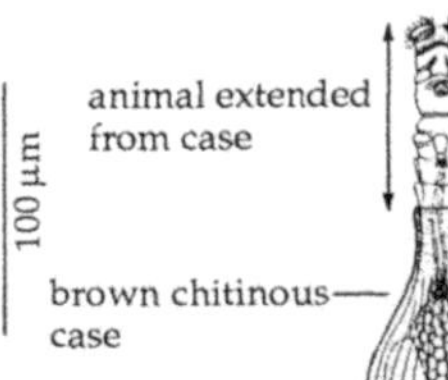

V. 5 *Habrotrocha angusticollis* (after Ward & Whipple, 1918)

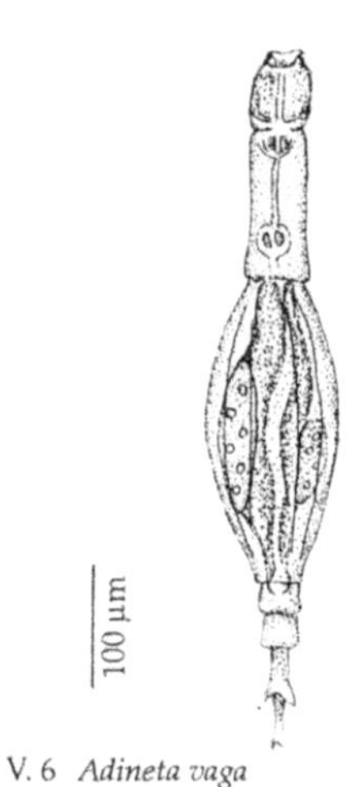

V. 6 *Adineta vaga*

6 Two red eye spots; wheel cilia short; length 200–400 µm *H. roeperi* (Milne)

– No eye spots; wheel cilia long; length up to 280 µm *H. reclusa* (Milne)

7 Snout or rostrum non-retractile; quick gliding movements. Common in *Sphagnum* *Adineta* (V.6)

– Snout or rostrum retractile 8

8 Toes (visible only when the body is fully extended) bearing cup-like suckers or joined to form a broad disc *Mniobia* (V.1)

– Toes not like this 9

9 Four-toed 10

– Three-toed 11

10 Spurs above the toes very long *Dissotrocha* (V.7)

– Spurs not very long *Philodina* (V.8)

11 Two red eyes usually present near end of long snout or rostrum; never carying an egg (young borne alive); usually long animals. Several species common in *Sphagnum* *Rotaria* (V.9)

– Eyes always absent; sometimes found carrying an egg. Some species have long spines *Macrotrachela*

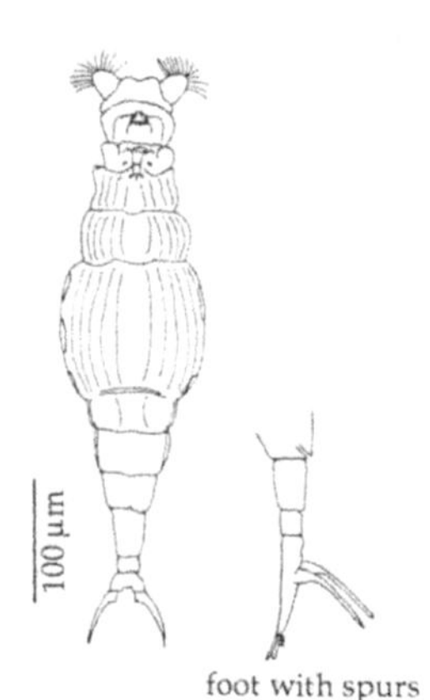

V. 7 *Dissotrocha aculeata* (after Murray, 1908b)

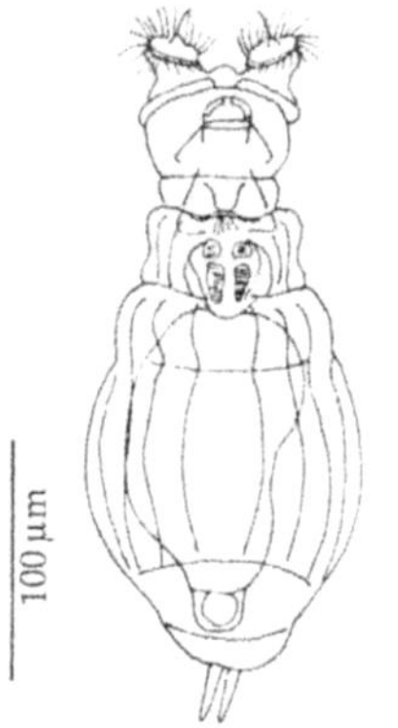

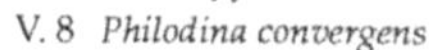
V. 8 *Philodina convergens*

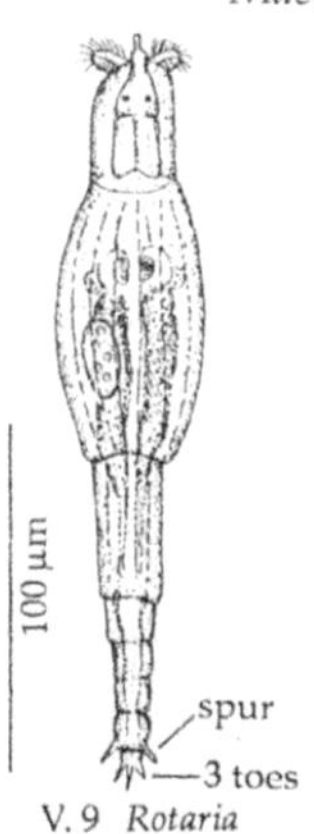

V. 9 *Rotaria*

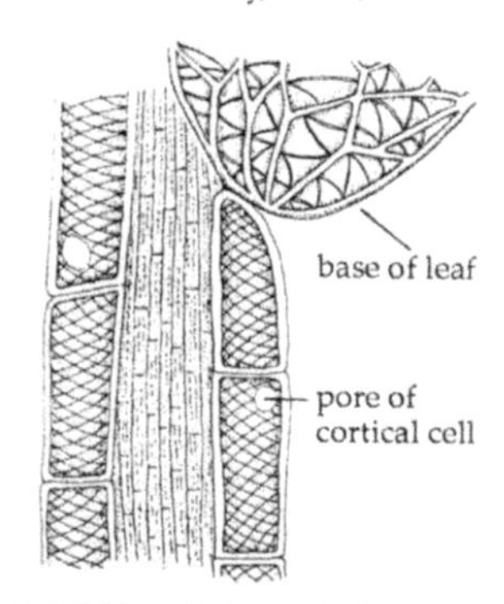

GII. 1 Part of a branch of *Sphagnum palustre*, showing spiral thickening of cortical cells.

Guide II. Some *Sphagnum* species common In Britain

Complete keys to the genus are given for Britain and Ireland by Smith (1978) and for Europe by Daniels & Eddy (1990). The key to British species by Hill and others (1992) is written for non-specialists. The following descriptions are based on simple field and microscopic characters. Some species are extremely variable especially in colour. Earlier or alternative names are given in brackets.

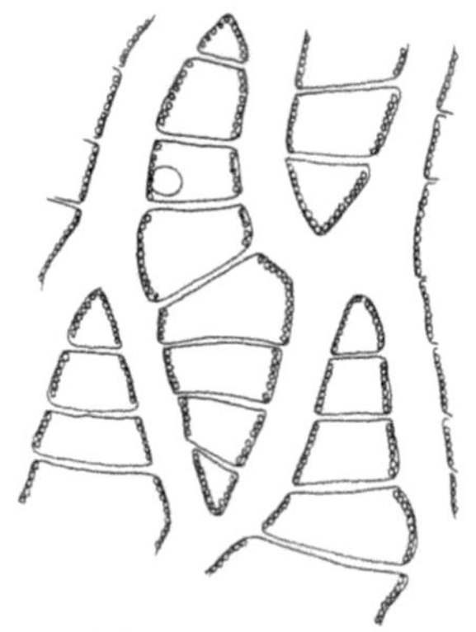

GII. 2 Part of leaf of *S. papillosum*, showing papillate walls of hyaline cells.

GII. 3 *S. squarrosum* (a) part of stem and group of branches showing spiky appearance, (b) branch leaves (after Daniels & Eddy, 1990).

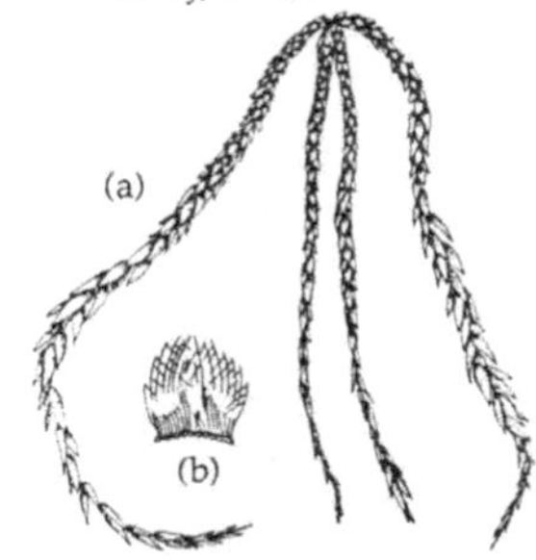

GII. 4 *S. fimbriatum* (a) branches showing feathery appearance, (b) stem leaf showing tattered edge (after Daniels & Eddy, 1990).

The first three species may be distinguished by the presence of fibrils in the branch cortex. These are easily seen if a small portion of the branch is placed on a slide in a drop of water and viewed under the low power of a compound microscope (GII.1).

S. papillosum Lindberg (pl. 1.4) is easily recognised by its characteristic ochre colour and its stubby branches. It may be confused with *S. palustre* in the field but if the leaves are examined under high power, the internal walls of the hyaline cells of the leaf are seen to be pimpled or papillose (GII.2). The plant forms green, yellowish or ochre tussocks on moors or in bogs. It is often hummock-forming. Confined to strongly acid localities, it is abundant in the north and west but local in the south-east.

S. palustre Linnaeus (*S. cymbifolium, S. latifolium, S. obtusifolium, S. centrale, S. subbicolor*) may be green, pinkish, orange or slightly yellowish. It forms tussocks on mesotrophic (pH 5.7–6.6) mires, streamsides and wet woodlands. It is abundant in the north and west and frequent in the south-east.

S. magellanicum Bride (pl. 1.1) is dull crimson, sometimes green or tinged with green. It forms reddish hummocks and tussocks often with *S. papillosum*. It can be distinguished from *S. palustre* by immersing a small piece of each in a weak alkali such as washing soda. *S. magellanicum* turns mud-brown; *S. palustre* shows little change of colour. *S. magellanicum* is sometimes abundant on well-developed convex bogs. It is local on open moors and valley bogs. Found throughout the British Isles but rare in the south-east.

S. squarrosum Crome is easily recognised by the leaf shape from which the species takes its name (GII.3). Most of the branch leaves are bent back sharply in mid-leaf giving a somewhat spiky appearance to the plant. It forms green 'lawns' amongst flowering plants and is sometimes abundant in swampy woodland. In smaller quantities it may be found in upland flushes which are areas of soil supplied by nutrient-rich water. It is confined to markedly eutrophic localities (pH above 6.5). It is frequent to common throughout the British Isles.

S. fimbriatum Wilson is always green with a large terminal bud and a somewhat feathery appearance (GII.4). It can easily be identified by pulling off the capitulum (bud cluster at top of stem), when the tattered or fimbriate stem leaves give a ruff-like appearance at the top of the stem. It forms extensive lawns in damp woods throughout the British Isles.

S. capillifolium (Ehrhart) Hedwig (*S. acutifolium, S. capillaceum, S. rubellum*) (pl. 1.2) is crimson, rose-pink or green with pink flecks. The stem leaves vary in shape and usually have conspicuous fibrils. It may form tussocks or carpets under heather. Characters used to separate this species from *S. subnitens* are given below. It is found in bogs and marshes and on moors which are not too wet or shaded. It is abundant in the north and west and frequent in bogs in the south-east.

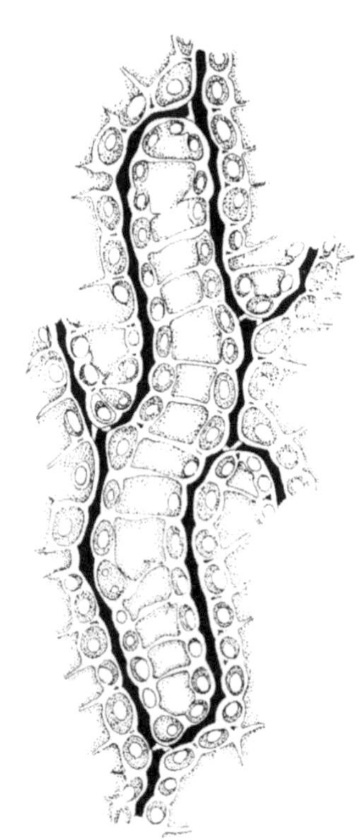
GII. 5 *S. auriculatum*, part of dorsal side of branch leaf showing pore distribution (after Daniels & Eddy, 1990).

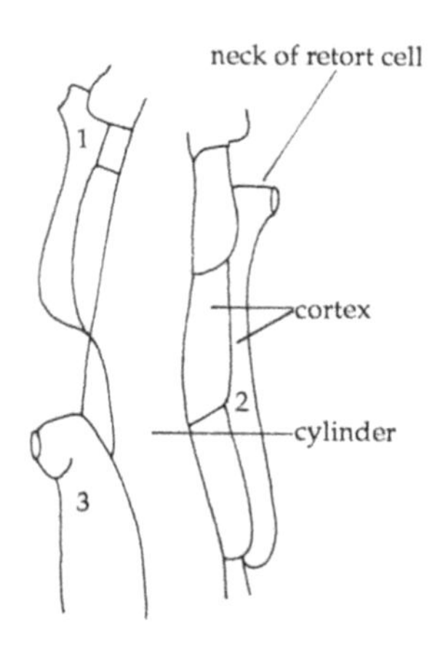

GII. 6 *S. tenellum*, part of branch showing 3 retort cells (1–3) (leaves removed).

S. subnitens Russow and Warnstorf (*S. acutifolium* varieties *lucidum* and *subnitens, S. plumulosum*) may be difficult to distinguish from the previous species but differs from it in that the centre of the capitulum is seldom redder than the surrounding branches and is often paler. The plants are green, brown, red or pink. The red is often dingy and dry plants often show a marked lustre or sheen. It occurs on tussocks, cushions and 'lawns' on moors, heaths and rocky banks and in woods. It is abundant in the north and west and common in the midlands and south.

S. compactum de Candolle (*S. rigidum*) is usually easily recognised by its whitish green, yellow or bright ochre colour and its densely crowded shoots. It forms compact tussocks on wet heaths or blanket bogs and amongst rocks, and is often found on the drier, higher ground at the edge of valley bogs. Frequent to common throughout the British Isles.

S. auriculatum Schimper (*S. contortum, S. subsecundum, S. lescurii*) is a very variable plant which is green, yellow or coppery red. The pore distribution on the outer side of a branch leaf is characteristic of the species (GII.5). The stem and branch leaves have fibrils in the hyaline cells. The variety *auriculatum* may have them throughout the stem leaf, whilst in the variety *inundatum* less than 40% of the stem leaf has fibrils. Variety *auriculatum* occurs in lawns and tussocks in woods or on moors, and submerged in ponds. Variety *inundatum* is found in marshes, heathy depressions, fens and flushes. Both are common in the north and west and frequent to common in the south and east.

S. cuspidatum Hoffman (pl. 1.5) has long green leaves and is usually found in pools and runnels in acid bogs and on moors. Out of the water, the so-called 'drowned kitten' appearance usually makes it easily recognisable, but there is considerable variation in colour and appearance between plants from slightly different habitats. It occurs throughout the British Isles, being common in the north and west and local in the south and east.

S. tenellum (Bridel) Bridel (*S. cymbifolium* variety *tenellum, S. molluscum*) is green or faintly brown. It has short branches with ovate, concave leaves. Under the microscope the necks of the retort cells of the branch cortex are very distinctive (GII.6), sticking out further than they do in any other British species. It occurs as scattered stems amongst other *Sphagnum* species on wet heaths and bogs. In high rainfall areas it also forms tussocks on sheltered rocky banks. Common in the north and west and local in the south and east.

S. recurvum Beauvois (*S. flexuosum, S. intermedium, S. fallax*) (pl. 1.3) is a very variable plant. The most widely distributed and common variety is *mucronatum*. The shoots are green or orange and it can usually be recognised by its pale stem and widely spaced, triangular, hanging stem leaves. It forms green or orange 'lawns' on moors and in wet woods, marshes and bogs throughout Britain and Ireland. Very common and often abundant in the north and west; local in the south and east.

Lists of genera and species recorded from *Sphagnum* in Britain, arranged alphabetically

* closely associated with *Sphagnum*

1 Desmids

Compiled with the help of A. Brook.

*Actinotaenium cucurbitinum, A. curtum; Arthrodesmus tenuissimum; Bambusina brebissonii; Closterium acutum, C. dianae, C. didymotocum, C. gracile, C. lunula, C. praelongum, C. pronum, C. rostratum, C. setaceum; Cosmarium amoenum, C. bioculatum, C. blytii, C. boekii, C. botrytis, C. brebissonii, C. concinnum, C. cucurbitum, C. cymatonotophorum, C. decedens, C. didymotocum, C. difficile, C. exiguum, C. formulosum, C. margaritiferum, C. moniliforme, C. ornatum, C. orthostichum, C. ovale, C. palangula, C. perpusillum, C. phaseolus, C. pseudoarctum, C. pseudoconnatum, C. punctulatum, *C. pygmaeum, C. pyramidatum, C. quadratum, *C. ralfsii, C. regnesi, C. scoticum, C. sphaerodeum, *C. sphagnicolum, C. subtumidum, C. tetraophthalmum, C. thwatesii, C. truncatellum, C. venustum; Cosmocladium constrictum; *Cylindrocystis brebissonii, *C. crassa, C. diplospora; Desmidium* species; *Euastrum ampullaceum, E. ansatum, E. bidentatum, E. binale, E. crassum, E. cuneatum, E. didelta, E. dubium, E. insigne, E. oblongum; Gonatozygon kinhani; Hyalotheca dissiliens; *Mesotaenium endlichearum, M. macrococcum; Micrasterias denticulata, *M. jenneri, *M. oscitans, M. purpureum, M. rotata, M. thomasiana, M. truncata; *Netrium digitus, N. oblongum; Penium cylindrus, P. nova-sylvae; Pleurotaenium minutum; Roya obtusa; Spirotaenia condensata, S. parvula; Spondylosium bambusoides, S. planum; Staurastrum dilatatum, S. margaritaceum, S. muricatum, S. ophiura, S. punctulatum, S. simonyii, S. teliferum; Staurodesmus dickei, S. glabrus, S. triangularis; *Tetmemorus brebissonii, T. granulatus; *Xanthidium antilopeum, X. armatum, X. orcadense, X. smithii, X. variabile.*

2 Diatoms

Compiled with the help of J. Carter, E. Haworth and D. Mann.
The taxonomy of this group is in a state of flux. The most up to date nomenclature has been used as far as possible.

*Achnanthes minutissima; Brachysira brebissonii, B. serians; Calvinula variostriata; *Eunotia alpina, *E. arcus *E. exigua, *E. glacialis, *E. incisa, *E. microcephala, *E. parallela, *E. pectinalis, *E. serra, *E. subtilissima, *E. tenella, *E. trinacria; Fragilariforma virescens; *Frustulia rhomboides; Navicula fossalis, N. hassiaca, *N. subtilissima; *Peronia fibula; *Pinnularia acoricola, *P. irrorata, *P. microstauron, *P. streptoraphe, P. subcapitata* var. *hilseana; Stauroneis; Stenopterobia sigmatella; Tabellaria flocculosa.*

3 Algae (excluding desmids, diatoms and flagellates)

Compiled from Pascher (1913–36), Pentecost (1984), Prescott (1951), Round (1981) and West & Fritsch (1927).
u, unicell; c, group or colony of cells; f, filament;
ep, epiphyte; en, endophyte.

Green algae (Chlorophyceae)

c-*Ankistrodesmus*; c-**Asterococcus superbus*; f-*Binuclearia*; f-*Bulbochaete*; f-*Cylindrocapsa*; u-**Desmatractum bipyramidatum*, u-*D. elongatum*; ep-*Dicranochaete brittanica*; f-*Draparnaldia*; u-**Eremosphaera viridis*; f-**Geminella minor*, f-*G. mutabilis*; c-*Gloeocystis vesiculosa*; f-*Microspora floccosa*; f-*Mougeotia*; c-**Nephrocytium agardhianum*; f-*Nitella*; ep-**Octogoniella sphagnicola*; f-*Oedogonium*; c-*Oocystis borgei*, u-*O. solitaria*; c-*Pediastrum boryanum*; ep-*Peroniella*; ep & en-**Phyllobium sphagnicola*; c-*Quadrigula*; c-*Scenedesmus obliquus*; c-*Schizochlamys delicatula*; f-*Spirogyra*; f-*Stigeoclonium*; f-*Zygnema*; f-**Zygogonium ericetorum* .

Yellow-green algae (Xanthophyceae)

Some epiphytes and endophytes classified as Xanthophyceae by botanists have been included with the plant-like flagellates or Phytomastigophora below.

c-**Chlorobotrys polychloris*, c-**C. regularis*; u-**Ophiocytium capitatum*; ep-**Perone dimorpha*; ep & en-*Rhizolekane*.

Yellow-brown algae (Chrysophyceae)

c-**Chrysostephanosphaera globulifera* (British status in doubt); c-**Phaeosphaera gelatinosa*.

4 Cyanobacteria (blue-green algae)

Symbols and sources are as for Algae (see **3** above).

f-*Anabaena*; f-*Chloroglea*; c-*Coccochloris*; c-*Cynarcus*; c-*Dactylococcopsis*; en-**Dermocarpa sphagnicola*; c-**Gloeocapsa rupestris*, c-**G. turgida*; f-*Gloeotrichia*; c-*Gomphosphaeria*; f-*Hapalosiphon hibernicus*, f-**H. intricatus*; f-*Hydrocoelus*; f-*Leptobasis*; c-*Merismopaedia punctata*; c-*Microcystis aeruginosa*; f-*Nostoc paludosum*, *Nostoc* species; f-*Oscillatoria*; f-*Phormidium*; f-*Scytonema*; u-**Synechococcus aeruginosus*; f-*Tolypothrix*.

5 Flagellates (Mastigophora)

Compiled from Gojdics (1953), Hall (1953), Leedale (1967), Pentecost (1984), Taylor (1987) and other sources.
Symbols as for Algae (see **3** above).

Green flagellates

*Carteria globosa; *Chlamydomonas acidophila, *C. sphagnicola; Euglena acus, E. deses,*E. mutabilis, E. oxyuris, E. pisciformis, E. sanguinea, E. spirogyra, E. tripteris, E. viridis; c-Gonium sociale; Gonyostomum semen; Hyalogonium klebsii; Lepocinclis; c-Platydorina; Polytoma uvella; Spermatozopsis; Trachelomonas aculeata, T. bulla, T. hispida.*

Yellow-green flagellates

en-**Heliochrysis sphagnicola*; en-**Myxochloris sphagnicola*; ep-**Stephanoporos sphagnicola*; ep-**S. scherffelii.*

Yellow-brown flagellates

Synura sphagnicola.

Armoured flagellates or dinoflagellates (mostly coloured)

Amphidinium; Ceratium hirundinella; **Cystodinium conchaeforme*; ep-*Dinococcales* (group of 10 species, attached to *Sphagnum* or other plants); *Glenodinium; Gymnodinium caudatum; Gyrodinium; *Hemidinium ochraceum; Katodinium stigmatica, K. vorticella; Peridinium cinctum, P. inconspicuum, P. limbatum, P. umbonatum, P. volzii, P. willei; Sphaerodinium; Woloszynskia.*

Colourless flagellates

Ancryomonas contorta; Astasia longa; Bodo parvus, B. saltans; Distigma proteus; Dinema sulcatum, D. entosiphon, D. mastigamoeba, D. mastigella; Notoselenus apocamptus; Oikomonas termo; Peranema trichophorum; Pleuromonas jaculans.

6 Naked amoebae

An incomplete list compiled from Cash, Wailes & Hopkinson (1905–21) and other sources.

*Amoeba proteus; Biomyxa vagans; *Chlamydomyxa labyrinthuloides; Naegleria gruberi; Pelomyxa palustris; Thecamoeba verrucosa.*

7 Sun animals or heliozoans

Compiled from Rainer (1968) and other sources.
z, species known to contain zoochlorellae.

Acanthocystis aculeata, A.erinaceus, A. pectinata, z-*A. penardi,* z-*A. turfaceae; Actinophrys sol; Actinosphaerium eichhorni; Chlamydaster sterni; Clathurina einkowskyi, C. elegans; Heterophrys fockei, H. myriopoda; Lithocolla globosa; Piniaciophora stammeri; Pompholyxophrys exigua, P. ovuligera; Raphidocystis glutinosa, R. tubifera; Raphidiophrys ambigua, R. intermedia.*

8 Testate rhizopods

Compiled from Cash, Wailes & Hopkinson (1905-21) and Corbet (1973).
*species commonly associated with *Sphagnum* and described by Corbet; z, known to contain zoochlorellae.

z-**Amphitrema flavum*, z-**A. stenostoma*, z-**A. wrightianum*; **Assulina muscorum*, **A. seminulum*; **Arcella discoides*, **A. gibbosa*, **A. hemisphaerica*, *A. mitrata*, *A. polypora*, **A. vulgaris*; **Bullinularia indica*; *Campascus minutus*; **Centropyxis aculeata* group, **C. arcelloides*, **C. cassis*; **Corythion dubium*, *C. pulchellum*; *Cryptodifflugia compressa*, *C. eboracensis*, *C. ovalis*, *C. oviformis*, *C. penardi*, *C. pulex*; *Difflugia amphoralis*, **D. bacilliarum*, **D. bacillifera*, *D. constricta*, *D. curvicaulis*, *D. globulus*, **D. oblonga*, **D. rubescens*, **D. tuberculata*, **D. urceolata*; **Euglypha acanthophora*, *E. brachiata*, **E. ciliata*, *E. cristata*, *E. filifera*, **E. rotunda*, *E. scutigera*, **E. strigosa*, **E. tuberculata*; *Heleopera lata*, **H. petricola*, **H. rosea*, z-**H. sphagni*, **H. sylvatica*; *Hyalosphenia cuneata*, z-**H. elegans*, *H. minuta*, *H. ovalis*, z-**H. papilio*; *Lecythium hyalinum*, *L. mutabile*; *Lesquereusia epistomium*, *L. inaequalis*, **L. modesta*, **L. spiralis*; **Nebela barbata*, **N. bigibbosa*, **N. carinata*, **N. collaris*, **N. dentistoma*, **N. flabellulum*, **N. galeata*, **N. griseola*, **N. lageniformis*, **N. marginata*, **N. militaris*, **N. minor*, **N. parvula*, **N. penardiana*, **N. tincta*, **N. tubulosa*, **N. vitraea*; *Phryganella acropodia*; *Placocista jurassica*, **P. spinosa*; *Portigulasia rhumbleri*; *Pseudochlamys patella*; **Quadrulella symmetrica*; *Pseudodifflugia compressa*; *Pyxidicula cymbalum*; *Sphenoderia dentata*, *S. fissirostris*, **S. lenta*, *S. macrolepis*; **Trigonopyxis arcula*; **Trinema enchelys*.

9 Ciliates

Compiled from Kahl (1930–35), Grolière (1975, 1977) and other sources and with help from Colin Curds.
*closely associated with *Sphagnum*; z, known to contain zoochlorellae.
Although many of these records are from mainland Europe, most ciliates are cosmopolitan in distribution and many probably occur amongst *Sphagnum* in Britain.

Blepharisma lateritium, *B. steini*, *B. musculus*, **B. sphagni*; *Bryometopus pseudochilodon*, z-**B. sphagni*; *Bryophyllum armatum*, *B. penardi*, *B. vorax*; *Bursaria truncatella*; *Chaenea*; *Chilodonella bavariensis*, *C. cucullus*, *C. uncinata*; z-*Climacostomum virens*; *Coleps*; *Colpidium*; *Colpoda*; **Cyclidium glaucoma*, z-**C. sphagnetorum*; *Cyclogramma protectissima*; *Cyrtolophosis mucicola*; *Dileptus tenuis*; *Drepanomonas dentata*, *D. exigua*, **D. sphagni*; *Enchelyodon ovum*, **E. sphagni*; *Euplotes patella*; *Frontonia*; *Gonostomum affine*; *Halteria grandinella*; **Hemicyclostyla sphagni*; **Histriculus sphagni*; z-*Holophrya*; *Keronopsis monilata*, *K. muscorum*, *K. wetzeli*; *Lacrymaria olor*; *Lembadion*; z-*Leptopharynx costatus*; *Litonotus fasciola*; **Malacophrys* *sphagni; Microthorax spiniger;* z-*Ophrydium versatile; Opisthotricha muscorum, O. parallela, O. sphagni; Oxytricha ludibunda, O. minor, O. variabilis; Parahistriculus minimus; Paraholosticha nana; Paramecium aurelia,* z-*P. bursaria; Pardileptus conicus; Perispira ovum;* z-*Platyophora viridis;* z-*Prorodon cinereus, P. gracilis, P. pyriforme; Pseudoblepharisma crassum; Psilotrocha teres; Pyxidium urceolatum; Sathrophilus havassei, S. vernalis; Spathidium amphoriforme, S. lionotiforme, S. muscicola; Spirostomum ambiguum, S. minus; Stentor coeruleus; Stichotricha aculeata; Stylonichia;* z-*Thylacidium truncatum; Trachelius;* **Trachelophyllum sphagnetorum;* **Trichopelma sphagnetorum; Uroleptus longicaudatus; Urostyla caudata;* z-*Urotricha agilis, U. ovata; Urozona buetschlii; Vaginicola; Vasciola picta;* z-*Vorticella muralis.*

10 Rotifers recorded from *Sphagnum* and bog pools in Britain

Compiled from Martin (1976, 1977) and from records in the British Museum of Natural History with the help of C. Hussey
*closely associated with *Sphagnum*; s, growing attached to *Sphagnum*.

Adineta vaga, A. gracilis, A. barbata; Aspelta circinator; Brachionus urceolaris; **Bryceella stylata,* **B. tenella, B. voigti; Cephalodella apocoela, C. auriculata, C. catellina, C. forficula, C. gibba, C. intuta, C. nana, C. pheloma, C. physalis, C. rostrum, C. tantilla, C. tantilloides, C. ventripes; Ceratrocha cornigera;* s-*Collotheca ambigua,* s-*C. annulata,* s-*C. calva,* s-*C. campanulata,* s-*C. coronetta,* s-*C. hoodii,* s-*C. ornata,* s-*C. quadrinodosa,* s-*C. spinata,* s-*C. trilobata; Colurella adriatica, C. obtusa, C. paludosa, C. tessellata;* s-*Conochilus; Dicranophorus hercules, D. longidactylum, D. lutkeni, D. robustus, D. rostratus, D. uncinatus; Dipleuchanis paludosa, D. propatula; Dissotrocha aculeata, D. macrostyla, D. spinosa;* **Elosa woralli; Encentrum felis, E. glaucum, E. mustela; Euchlanis incisa, E. meneta, E. parva, E. proxima, E. triquetra; Filinia terminalis;* s-*Floscularia conifera; Gastropus hyptopus, G. minor;* **Habrotrocha angusticollis, H. bidens, H. collaris, H. constricta, H. elegans,* **H. lata, H. longula, H. milnei, H. minuta,* **H. reclusa,* **H. roeperi; Keratella quadrata,* **K. serrulata;* **Lecane agilis, L. clara, L. closterocerca, L. cornuta, L. depressa, L. flexilis, L. galeata, L. hamata, L. inermis, L. lunaris, L. ploenensis, L. pyrrha, L. signifera, L. stichaea; Lepadella acuminata, L. ovalis, L. patella, L. pterygoides, L. triba, L. triptera; Lindia torulosa; Macrochaetus collinsi; Macrotrachela concinna, M. multispinosa, M. papillosa, M. plicata, M. quadricornifera; Microdon clavus; Mniobia incrassata, M. magna, M. symbiotica; Monommata acticis, M. aeschyna, M. astia, M. longiseta, M. maculata, M. phoxa; Mytilina mucronata, M. ventralis* var. *brevispina; Notommata allantois, N. cerebrus, N. contorta, N. copeus, N. falcinella, N. groenlandica, N. pachyura, N. pavida, N. saccigera, N. tripus; Philodina acuticornis, P. brevipes, P. nemoralis, P. rugosa; Ploesoma lynceus;* **Polyarthra minor, P. vulgaris; Proales decipiens, P. doliaris, P. fallaciosa, P. latrunculus, P. micropus,*

P. minima; Proalinopsis caudatus, P. squamipes; s-*Ptygura brachiata,* s-**P. longicornis,* s-*P. longipes,* s-*P. pilula,* s-*P. rotifer,* s-**P. velata; Resticula melandocus, R. nyssa; Rotaria haptica, R. macrura, R. magna-calcarata, R. neptunoida, R. quadrioculata, R. socialis, R. spicata, R. tardigrada; Scepanotrocha rubra; Squatinella longispinata, S. microdactyla, S. mutica, S. tridentata;* s-*Stephanoceros fimbriatus, S. millsii; Streptognatha lepta; Synchaeta pectinata; Taphrocampa annulosa, T. clavigera; Testudinella emarginula, T. patina; Tetrasiphon hydrocora; Trichocerca bicristata, T. cavia, T. collaris, T. elongata, T. longiseta, T. porcellus, T. rattus, T. rosea, T. junctipes, T. tigris; Trichotria pocillum, T. tetractis, T. truncata; Wierzejskiella velox.*

11 Flatworms (Microturbellaria)

Compiled from Young (1970) and other sources. Apart from *C. sphagnetorum* only species on Maitland's (1977) checklist have been included. Other species may occur.
*closely associated with *Sphagnum*; z, known to contain zoochlorellae.

z-*Castrada armata;* z-**C. sphagnetorum; Castrella truncata; Catenula lemnae; Geocentrophora sphyrocephala, G. baltica; Gyratrix hermaphroditus; Microdalyiella kupelwieseri; Opistomum pallidum; Prorhynchus stagnalis; Rhynchomesostoma rostratum; Stenostomum leucops, S. unicolor;* z-*Typhloplana viridata.*

12 Nematode worms

Compiled from Goodey (1951), Goodey & Goodey (1963) and other sources.
M, in Maitland's (1977) checklist of freshwater species. Others may be terrestrial or not yet recorded from Britain.

M-*Alaimus primitivus;* M-*Aglenchus agricola;* M-*Amphidelus dolichurus;* M-*Anaplectus granulosus;* M-*Aphelenchoides heliophilus,* M-*A. parietinus;* M-*Bunonema reticulatum; Criconema; Criconemoides sphagni; Diplogaster sphagni; Dorylaimus;* M-*Eucephalobus elongatus;* M-*Eudorylaimus carteri;* M-*Euteratocephalus crassidens,* M-*E. palustris; Hemicycliophora membranifera;* M-*Hirschmannia gracilis;* M-*Ironus ignavus;* M-*Monohystera filiformis,* M-*M. vulgaris;* M-*Plectus cirratus,* M-*P. parietinus,* M-*P. parvus,* M-*P. rhizophilus;* M-*Prismatolaimus intermedius;* M-*Prodorylaimus longicaudatus; Rhabdites uliginosa;* M-*Rotylenchus robustus;* M-*Teratocephalus terrestris;* M-*Tripyla monohystera, T. papillata, T. pellucida; Tylencholaimus;* M-*Tylenchus davainei,* M-*T. filiformis; Wilsonema otophorum.*

13 Segmented worms

Compiled from Springett (1970), Standen & Latter (1979), Griffiths (1973) and other sources.
*closely associated with *Sphagnum*; M, in Maitland's (1977) checklist of freshwater species.

M-*Aeolosoma hemprici; Bimastos eiseni; Cernosvitoviella briganta*; M-*Cognettia cognetti*, M-*C. glandulosa*, M-**C. sphagnetorum*; *Marionina clavata*; M-*Pristina idrensis*; M-*Vejdovskyella comata*.

14 Gastrotrichs

Compiled from Martin (1981). All these species are from bog pools at Thursley Common, Surrey. Martin also records several new species not listed below.

Chaetonotus heterocanthus, M-*C. maximus*, *C. ophiogaster*, *C. polyspinosus*, *C. voigti*, M-*C. zelinkai*; *Heterolepidoderma ocellatum*; *Ichthydium forcipatum*; M-*Lepidodermella squamatum*; *Stylochaeta fusiformis*.

15 Tardigrades

From Harnisch (1929).

Hypsibius scoticus; Macrobiotus species.

16 Crustaceans

Compiled from Harding & Smith (1974), Pearsall (1950), Scourfield & Harding (1958) and other sources.

Cladocerans
Acantholeberis curvirostris; Acroperus harpae, A.rustica; Ceriodaphnia setosa; Chydorus sphaericus group; *Graptoleberis testudinaria; Rhynchotalona rostrata; Scapholeberis mucronata; Simocephalus vetulus; Streblocerus serricaudatus*.

Copepods
**Acanthocyclops venustus; Bryocamptus weberi; *Moraria sphagnicola*.

17 Oribatid mites recorded from *Sphagnum* in Britain

Compiled with help from Malcolm Luxton.

*Achiptera nitens; Adamaeus (Damaeus) onustus; Adoristes ovatus; Autogneta longilamellata; Banksinoma lanceolata; Camisia spinifer; Carabodes labyrinthicus, C. marginatus; Cepheus latus; Ceratoppia bipilis; Ceratozetes gracilis; Chamobates cuspidatus, C. schuetzi; Diapterobates humeralis; Dissorhina ornata; Edwardzetes edwardsi; Eupelops farinosus; Euzetes nitens; Fuscozetes fuscipes; Galumna lanceata; Hermannia gibba, H. reticulata; Hermaniella picea; Hoplophthiracarus pavidus; *Hydrozetes lacustris; Hypochthonius rufulus; *Limnozetes ciliatus; Liochthonius brevis; Malaconothrus monodactylus; Melanozetes mollicomus,*

**M. stagnatilis; Minunthozetes semirufus; Moritzoppia unicarinata; *Mucronothrus nasalis; Mycobates sarakensis; *Nanhermannia nana; Nothrus coronata, N. palustris, N. pratensis, N. silvestris; Odontocepheus elongatus; Opiella nova; Oribatula tibialis; Parachipteria punctata; Peloptulus montanus, P. phaeonotus; Phauloppia lucorum; Phthiracarus affinis; Platynothrus peltifer, P. punctatus, P. thori; Quadroppia quadricarinata; Rhysotritia ardua, R. duplicata; Sphaerozetes piriformis; Steganacarus magnus; Suctobelba trigona; Tectocepheus velatus; *Trhypochthoniellus crassus; *Trimalaconothrus glaber, T. maior, *T. tardus; Xenillus tegeocranus.*

18 Mites of other groups

Mesostigmata

Cilliba cassidea (pl. 8.2).
This species is very flattened and almost circular; sometimes found amongst woodland *Sphagnum*.

Hydracarina (water mites)

**Hydrovolzia placophora.*

Spiders and Insects are not listed here, but see Key I for references.

5 Techniques and approaches to original work

Collecting

A few medium-sized plastic bags, some labels, a notebook and a pencil are the modest requirements. It is wise not to go too far, or unaccompanied, into bogland. Footprints on the *Sphagnum* carpet take a long time to disappear. Only a small handful of each *Sphagnum* species is needed. It is less harmful to take a few stems from each of several different places, in the pool, the hummock, the lawn or the woodland fen, than to take one large handful from each place. Written permission must be obtained to collect in Nature Reserves. It is better to collect outside such areas, as sparingly as possible, and only where there is abundant growth of the moss.

Any time of the year is suitable for collecting; different organisms are likely to be found at different seasons.

Material from different sites is put into separate labelled polythene bags, which are closed so that the moss does not dry up.The contents of the bags should be given a preliminary examination as soon as possible. Delicate organisms, such as ciliates, flagellates and rotifers should be examined first. Bags may be kept in cool, light conditions and loosely tied so that other organisms may be examined in more detail when convenient. Submerged species like *Sphagnum cuspidatum* may be kept out of doors in a plastic pail with a little rain water. Even many months later they will yield interesting material.

The composition of the *Sphagnum* community changes rapidly in response to changes in the temperature, light intensity, humidity and pH in which it is kept.

Microscopes

For detailed studies and accurate identification of species, a well-maintained compound microscope magnifying up to 400x is essential. An oil immersion lens and phase contrast system are desirable extras. A mechanical stage makes it easier to count organisms and a binocular head helps prevent fatigue. It is sometimes possible to borrow a microscope from a local school or college during school holidays.

Making temporary slides

Two or three spreading branches of the moss may be mounted on a microscope slide in a drop of water. Always use a coverslip when you examine wet material under a compound microscope; otherwise the objective of the microscope is likely to get wet. The coverslip should be lowered onto the preparation carefully to avoid trapping air bubbles. Excess water should be soaked up with filter paper or a tissue. Hanging or pendent branches and a small piece of the main stem may be similarly mounted. It may be necessary to strip off leaves from a branch with forceps to examine them separately or to see the retort cells of the branch cortex more clearly.

Making 'squeezes' is a simple and very useful technique. For a quick look or preliminary examination, a few drops of water are squeezed from damp or wet bog moss onto a microscope slide and covered with a coverslip. Dry moss should be soaked in clean water for a few minutes before squeezing.

For a more comprehensive sample of the organisms in the water film clinging to the plants and from inside the cells, the 'squeeze and soak' technique may be used. A few moss plants are squeezed into a petri dish or similar container, soaked for a few minutes in clean water and squeezed again into the dish. This process may be repeated a standard number of times to make comparisons between samples.

From the dish it is possible, with the aid of a hand lens, to pick out larger organisms such as insects, mites and worms, using a fine paint brush or strip of filter paper to tranfer them quickly to a drop of clean water on a slide. To sample smaller organisms, the contents of the dish are stirred well and a drop of water is removed with a pipette to a slide for examination under the microscope.

Some special methods for flagellates, ciliates and rotifers

The free-swimming behaviour of flagellates and ciliates may be observed by using a cavity slide, or an ordinary plane slide with the coverslip supported at each corner by a small blob of vaseline or plasticine. When an ordinary slide is used with a supported coverslip, evaporation of water during observation causes larger organisms to slow down, and eventually distorts and kills them. They may be observed more easily when slowed down by a tangle of debris or leaves or even strands of cotton wool. Viscous liquids such as methyl cellulose (Methocell) or Polycell water paste can be used to slow these organisms down by placing the viscous liquid in a ring around the drop of water containing the organisms, and then applying a coverslip. The paste is made by soaking 1 g of Polycell in 45 ml water, boiling for about 30 minutes and then mixing with 45 ml cold water.

Other slowing agents include 2% nickel sulphate solution, 1% copper sulphate solution or 4% formalin (**toxic**). These may be introduced by placing a drop near the edge of the coverslip and drawing it through slowly with a piece of filter paper from the other side. Stains or preservatives may be introduced in a similar way; 0.5% methyl green in 1% acetic acid stains nuclei; Lugol's iodine (4 g iodine dissolved in 100 ml of a 6% solution of potassium iodide, stored in a brown bottle and diluted to the colour of weak tea when used) shows up cilia and flagella.

Rotifers may be killed in an extended position by placing a drop of water containing them in a small flat-bottomed dish with an equal volume of boiling water. To preserve them, 2.5% formalin solution (**toxic**) may be used, to which 2% glycerine and a small drop of eosin have been added. The red dye, eosin, stains the animals, making them easier to locate and study.

Critical lighting or phase contrast improve visibility of transparent organisms. Jahn, Bovee & Jahn (1979) and Finlay, Rogerson & Cowling (1988) give further methods for handling and examining protozoa and Pennak (1989) is useful for protozoa and other organisms.

A damp chamber for keeping organisms under observation

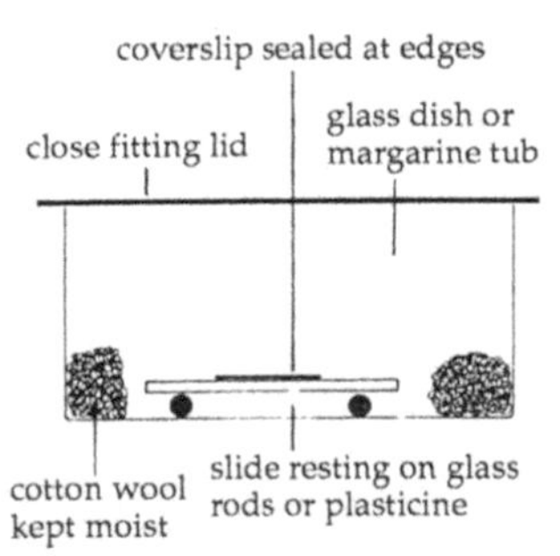

Fig. 55. Keeping slides of live material in a damp chamber (after Guthrie, 1989).

Sometimes it is necessary to observe organisms over a period of time. The drop of water containing the organisms on a slide may be ringed around with vaseline in such a way as to seal the edges of the coverslip when it is placed in position. The organisms can be studied for a number of hours without drying up. If placed in a damp chamber as shown in fig. 55, they may be kept for several days or longer. The inclusion of a few fresh *Sphagnum* leaves on the slide should ensure sufficient oxygen. Cavity slides are ideal for the purpose.

Making semi-permanent slides

Hoyer's medium can be used to kill, fix, clear and preserve many of the organisms found in *Sphagnum*. It is made by dissolving 30 g of clear crystals of gum arabic in 50 g cold deionised water, adding 200 g chloral hydrate (**toxic**), allowing this to dissolve completely, then adding 20 g glycerine. The solution should be used from a dropping bottle.

Moss leaves, branches and stems, larger organisms such as mites, or insects and their larvae can be mounted directly in the medium. If a drop of squeezed-out liquid from *Sphagnum* is allowed to dry slowly on a slide over a radiator or under a lamp, a smear is formed. When this is just dry, a drop of Hoyer's medium is added and a coverslip lowered carefully onto it. This method destroys chlorophyll and delicate organisms, but it is excellent for the cases of testate rhizopods and for other organisms with resistant outer coverings. After about a month the coverslip should be ringed with colourless nail varnish to seal it. When dry, the slides may be stored in slide boxes and kept for several years, labelled with the date, place of collection and other relevant information.

Measuring and counting organisms

Many of the organisms living in *Sphagnum* lose their characteristic appearance when they die, and the only way to keep permanent records is by drawing live specimens. Simple annotated line drawings are required, without any shading, and the relative size and shape of all recognisable features should be shown. In all cases accurate measurements must be added.

Objects seen under the microscope are measured in micrometres (µm). (A micrometre is 1/1000 of a millimetre.) Rough measurements can be made by comparing the length of the organism with the diameter of the field of view of the microscope. This may be measured with a low power objective x10 in place and x10 eyepiece, by viewing the millimetre scale on a plastic ruler. For measurements using high power objectives (x40 and above), a micrometer eyepiece should be calibrated against a micrometer slide (fig. 56). Guthrie (1989) gives details. Because of variability of size within a species, an accuracy of about 10% is adequate for identification of all but the smallest organisms. Sizes are recorded by adding to each drawing a scale line labelled with its length in micrometres.

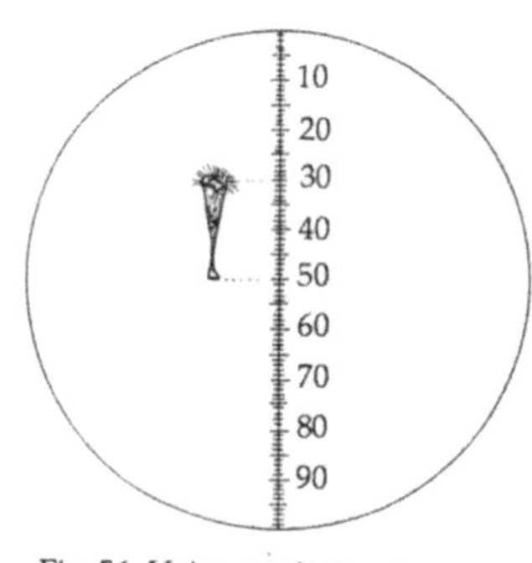

Fig. 56. Using a micrometer eyepiece (after Guthrie, 1989). This animal is 20 units long, and previous calibration with a micrometer slide showed that each eyepiece unit represents 16 µm. The animal's length is therefore 320 µm.

A Sedgewick-Rafter cell is a commercially available glass chamber holding 1 ml of liquid spread 1 mm deep over a large area. It can be used to count the organisms in 1 ml of water squeezed from *Sphagnum*. Alternatively, a known volume of water can be deposited on a slide and covered completely with a large coverslip so that all parts of the drop are clearly visible. A grid can be marked on the back of the slide with a felt tipped pen or a diamond knife. Counting is best done on a mechanical stage, so that the slide can be moved evenly from side to side or from front to back.

Estimating diversity

A habitat that is 'good' for, say, testate rhizopods has many species belonging to that group. One way to express this quality is to count species in a standard sample of moss or water. But the number of species you find will depend in part on the number of individuals you see. An index of diversity that takes into account the numbers of both species and individuals can be useful, and need not require naming of species or exact standardisation of sample size. Many types of diversity index are described by Magurran (1988). One of the most straightforward for use with *Sphagnum* organisms is the Sequential Comparison Index (SCI) (Cairns and others, 1968). It could be used, for example, to show

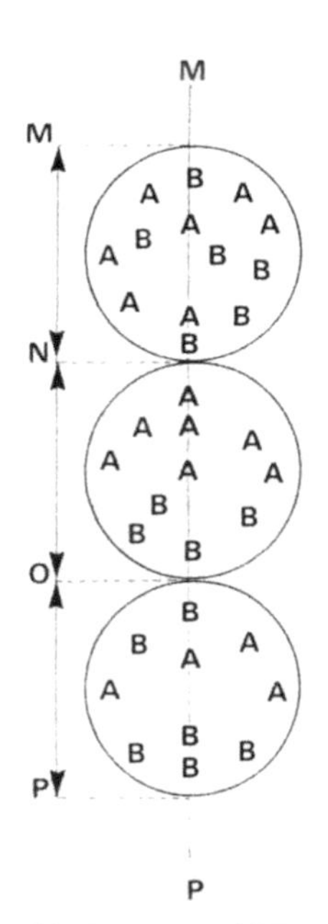

Fig. 57. Microtransect M–P across three adjacent fields of view, MN, NO and OP, in which organisms A and B are unevenly distributed.

how diversity of testate rhizopods changes with depth or pH, and to compare this profile with that for another group such as the desmids.

A micrometer eyepiece can be used to delimit a microtransect across the slide (fig. 57). Consecutive individuals along this microtransect are compared and a note is made of the number of times an individual is encountered which is of a different species from its immediate predecessor. Each sequence of consecutive identical individuals constitutes a run. The index is then calculated as:

SCI = number of runs/number of individuals, for example:

1 2 3 4 5 6 7 8 9 10 11 12
A A B A A A B B A A A A = 5 runs
SCI = $^5/_{12}$ = 0.42

1 2 3 4 5 6 7 8 9 10 11 12
A A A A A A A A A B B A = 3 runs
SCI = $^3/_{12}$ = 0.25

Measuring environmental factors in the field

The biology departments of most schools and colleges have instruments for measuring environmental factors such as light, temperature, oxygen concentration and electrical conductivity. These come with full instructions and are easy to use in the habitats in which *Sphagnum* grows for carrying out the investigations suggested in Chapter 2 above. Suppliers are listed on p. 63.

With these instruments it is possible to make measurements easily and precisely, but less expensive alternatives are available. Unwin & Corbet (1991) give instructions for building and using simple light meters and devices for measuring temperature. Oxygen concentrations can be estimated with the micro Winkler technique (Hingley, 1979) or with a field kit (suppliers p. 63). Universal test papers or simple kits supplied by gardening firms may also be used to measure pH.

Field experiments (fig. 58)

The diagrams below show home-made apparatus suitable for carrying out projects on colonisation by various types of organisms (see p. 10). Microscope slides are set out in standard positions in the water or moss, and left for periods ranging from a few days to several weeks before being brought into the laboratory for microscopic examination of the algae and other organisms growing on them.

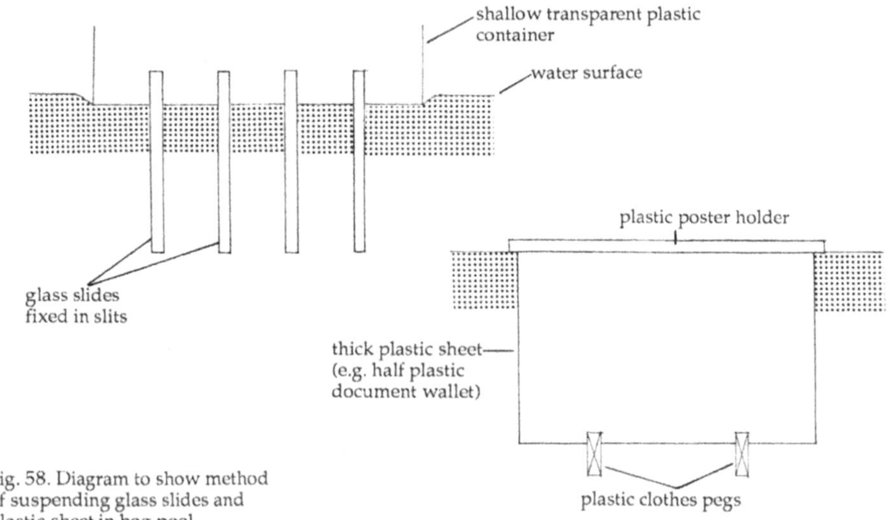

Fig. 58. Diagram to show method of suspending glass slides and plastic sheet in bog pool.

Presentation of results

Writing up is an important part of a research project. A really thorough, critical investigation that has established new information of general interest may be worth publishing, if the organisms mentioned can be reliably identified. Journals that might consider short papers on microscopic animals and plants in *Sphagnum* include the journals of local natural history societies, or *Microscopy, Field Studies, British Journal of Entomology and Natural History, School Science Review* and *Journal of Biological Education*. Examine recent numbers of these journals to see what sort of thing they publish, and then write a paper along similar lines, keeping it as short and concise as possible while still presenting enough information to establish the conclusions. Advice from an expert is helpful at this stage.

It is an unbreakable convention of scientific publication that results are reported with scrupulous honesty. It is therefore essential to keep detailed and accurate records throughout the investigation, and to distinguish in the write-up between certainty and probability, and between deduction and speculation.

It will often be necessary to apply statistical techniques to test the significance of the findings. A book such as the *Open University Project Guide* (Chalmers & Parker, 1989) will help, but expert advice can contribute much to the planning, as well as to the analysis, of the work.

Further reading

Finding books

Some of the books and journals listed here will be unavailable in local and school libraries. It is possible to make arrangements to see or to borrow such works by seeking permission to use the library of a local university, or by asking your local public library to borrow the work (or a photocopy of it) for you via the British Library, Document Supply Centre. This may take several weeks, and it is important to present your librarian with a reference that is correct in every detail. References are acceptable in the form given here, namely the author's name and date of publication, followed by (for a book) the title and publisher or (for a journal article) the title of the article, the journal title, the volume number and the first and last pages of the article.

Asterisks mark publications available from The Richmond Publishing Co. Ltd., P.O. Box 963, Slough, SL2 3RS.

Freshwater Biological Association Scientific Publications are available from the Freshwater Biological Association, The Ferry House, Far Sawrey, Ambleside, Cumbria, LA22 OLP.

References

Amoros, C. (1984). Crustacés Cladocères. *Bulletin mensuel de la Société Linnéenne de Lyon*, 53e année, nos. 3 & 4, pp. 72–145. (Available from Société Linnéenne de Lyon, 33 Rue Bossuet, 69003 Lyon, France.)

Barber, H.G. & Haworth, E.J. (1981). *A Guide to the Morphology of the Diatom Frustule*. Freshwater Biological Association Scientific Publications **44**.

Barnes, R.S.K. (ed.)(1984). *A Synoptic Classification of Living Organisms*. Oxford: Blackwell Scientific Publications.

Bartos, E. (1951). The Czechoslovakian Rotatoria of the order Bdelloidea. Vestník Ceskoslovenské Zoologické Spolecnosti **15**, 1–224.

Basilier, K., Granhall, V. & Stenstrom, T.A. (1978). Nitrogen fixation in the wet minerotrophic moss community of a sub-arctic mire. *Oikos* **31**, 236–246.

Belcher, H. & Swale, E. (1976). *A Beginner's Guide to Freshwater Algae*. London: HMSO.

Belcher, H. & Swale, E. (1979). *An Illustrated Guide to River Phytoplankton*. London: HMSO.

Brinkhurst, R.O. (1971). *A Guide for the Identification of British Aquatic Oligochaeta*. Freshwater Biological Association Scientific Publications **22**.

Brook, A.J. (1981). *The Biology of Desmids*. Oxford: Blackwell Scientific Publications.

Brook, A.J. & Lind, E.M.(1980). *Desmids of the English Lake District*. Freshwater Biological Association Scientific Publications **42**.

Brook, A.J. & Williamson, D.B. (1990). *Actinotaenium habeebense* (Irénée Marie) nov. comb., a rare drought-resistant desmid. *British Phycological Journal* **25**, 321–327.

Cairns, J., Albaugh, D.W.,Busey, F. & Channay, M.D.(1968). The sequential comparison index. *Journal of the Water Pollution Control Federation* **40**, 1607–1613.

Cash, J., Wailes G.H. & Hopkinson, J. (1905–1921). *The British Freshwater Rhizopoda and Heliozoa* (5 vols). London: Ray Society.

Clymo, R.S. (1963). Ion exchanges in *Sphagnum* in relation to bog ecology. *Annals of Botany* **27**, 309–324.

Corbet, S.A. (1973). Microscopic animals in *Sphagnum* moss: an illustrated introduction to the testate rhizopods around Malham Tarn, Yorkshire. *Field Studies* **3** (5), 801–838.

Cranston, P.S. (1982). *A Key to the British Orthocladiinae (Chironomidae)*. Freshwater Biological Association Scientific Publications **45**.

Crowe, J.H., Crowe, L. & Chapman, D. (1984). Preservation of membranes in anhydrobiotic organisms: the role of trehalose. *Science* **223**, 701–703.

*Curds, C.R. (1982). *British and other Freshwater Ciliated Protozoa, Part I. Ciliophora: Kinetofragminophora*. Synopses of the British Fauna **22**. Cambridge: Cambridge University Press.

*Curds, C.R., Gates, M.A. & Roberts, D.M. (1983). *British and other Freshwater Ciliated Protozoa, Part II. Ciliophora: Oligohymenophora and Polyhymenophora*. Synopses of the British Fauna **23**. Cambridge: Cambridge University Press.

Daniels, R.E. & Eddy, A. (1990). *Handbook of the European Sphagna*. Institute of Terrestrial Ecology. London: HMSO.

Donner, J. (1966). *Rotifers* (trans. H.G.S. Wright). London: Frederick Warne.

Fantham, H.B. & Porter, A. (1945). The microfauna, especially the protozoa, found in some Canadian mosses. *Proceedings of the Zoological Society of London* **115**, 97–174.

Finlay, B.J. & Fenchel, T. (1989). Everlasting picnic for protozoa. *New Scientist*, 1 July 1989.

Finlay, B.J., Rogerson, A. & Cowling, A.J. (1988). *A Beginner's Guide to the Collection, Isolation, Cultivation and Identification of Freshwater Protozoa*. Ambleside: Culture collection of algae and protozoa, Freshwater Biological Association.

Fjellberg, A. (1980). *Identification Keys to Norwegian Collembola*. Oslo: Norwegian Entomological Society.

*Friday, L.E.A. (1988). *A Key to the Adults of British Water Beetles*. Field Studies **7** (1), 1–151. An AIDGAP key.

Gilbert, J.J. (1974). Dormancy in rotifers. *Transactions of the American Microscopical Society* **93** (4), 490–513.

Gledhill, T. (1960). Some water mites (Hydrachnellae) from seepage water. *Journal of the Quekett Microscopical Club* Series, **4**, **5** (11), 293–307.

Gojdics, M. (1953). *The Genus* Euglena. Madison: University of Wisconsin Press.

Goodey, T. (1951). *Soil and Freshwater Nematodes*. London: Methuen.

Goodey, T. & Goodey, J.B. (1963). *Soil and Freshwater Nematodes*. New York: Wiley.

Granhall, V. & Hofsten, A. (1976). Nitrogenase activity in relation to intracellular organisms in *Sphagnum* mosses. *Physiologia Plantarum* **36**, 88–94.

Granhall, V. & Selander, H. (1973). Nitrogen fixation in a sub-arctic mire. *Oikos* **24**, 8-15.

Griffiths, D. (1973). The structure of an acid moorland pond community. *Journal of Animal Ecology* **42**, 263–284.

Grolière, C.-A. (1975). A description of some hypotrich ciliates from *Sphagnum* bogs and acid water ponds. *Protistologica* **11**, 481–498. (In French with English summary.)

Grolière, C.-A. (1977). A contribution to the study of ciliates from *Sphagnum* mosses. **II**. Dynamics of the populations. *Protistologica* **13**, 335–352. (In French with English summary.)

Grospietsch, T. (1965). *Wechseltierchen (Rhizopoden)*. Stuttgart: Kosmos, Franckh'sche Verlagshandlung.

*Guthrie, M. (1989). *Animals of the Surface Film*. Naturalists' Handbooks No. **12**. Slough: The Richmond Publishing Co. Ltd.

Hall, R.P. (1953). *Protozoology*. London: Prentice-Hall.

Harding, J.P. & Smith, W.A. (1974). *A Key to the British Freshwater Cyclopid and Calanoid Copepods*. Freshwater Biological Association Scientific Publications **18**.

Harnisch, O. (1929). Die Biologie der Moore. *Die Binnengewässer* VII: 1–146. Stuttgart: E. Schweizarbeit'sche Verlagsbuchhandlung.

Hayward, P.M. & Clymo, R.S. (1982). Profiles of water content and pore size in *Sphagnum* and peat, and their relation to peat bog ecology. *Proceedings of the Royal Society of London B* **215**, 299–325.

Henderson, P. A. (1990). *Freshwater ostracods.* Synopses of the British Fauna (New Series) No. 42. Oegstgeest, The Netherlands: Universal Book Services/ Dr W. Backhuys.
Hill, M.O., Hodgetts, N.G. & Payne, A.G. (1992). Sphagnum: *A Field Guide.* Peterborough: Joint Nature Conservation Committee. (Available from Natural History Book Service Ltd., 2 Wills Road, Totnes, Devon TQ9 5XN.)
Hingley, M.R. (1979). *Fieldwork Projects in Biology.* Poole, Dorset: Blandford Press.
Hudson, C.T. & Gosse, P.H. (1889). *The Rotifera; or Wheel-animalcules* (2 vols). London: Longmans Green.
Ingold, C.T. (1975). *An Illustrated Guide to Aquatic and Water-borne Hyphomycetes.* Freshwater Biological Association Scientific Publications **30**.
*Jahn, T.L., Bovee, E.C. & Jahn, F.F. (1979). *How to know the Protozoa* (2nd edn). Iowa: Brown & Co.
*Jones, D. (1983). *The Country Life Guide to Spiders of Britain and Northern Europe.* Feltham: Newnes.
Jones-Walters, L. (1989). Keys to the families of British Spiders. *Field Studies* **7**, 365–886.
Kahl, A. (1930-35). Urtiere oder Protozoa. I: Wimpertiere oder Ciliata (Infusoria). *Tierwelt Deutschlands* **18, 21, 25, 30**: 1–886.
Koste, W. (1978). *Rotatoria. Die Rädertiere Mitteleuropas* (2 vols). Berlin: Borntraeger. (A revision of Voigt (1956).)
Krammer, K. & Lange-Bertalot, H. (1986–). *Süsswasserflora von Mitteleuropa. Bacillariophyceae* (4 vols). Jena. Gustav Fischer.
Lee, J.J., Hutner, S.H . & Bovee, E.C. (ed.) (1985). *Illustrated Guide to the Protozoa.* Kansas, USA: Society of Protozoologists.
Leedale, G.F. (1967). *Euglenoid Flagellates.* Englewood Cliffs, New Jersey: Prentice-Hall.
Lind, E.M. & Brook, A.J. (1980). *Desmids of the English Lake District.* Freshwater Biological Association Scientific Publications **42**.
Magurran, A.E. (1988). *Ecological Diversity and its Measurement.* London: Croom Helm.
Maitland, P.S. (1977). *A Coded Checklist of Animals Occurring in Freshwater in the British Isles.* Institute of Terrestrial Ecology, Natural Environment Research Council.
Martin, L.V. (1976, 1977). Rotifers in *Sphagnum* pools on Thursley Common. *Microscopy* **33**, 90–93, 236–241.
Martin, L.V. (1981). Gastrotrichs found in Surrey. *Microscopy* **34**, 286–300.
Montet, G. (1915). Contribution a l'étude des Rotateurs du bassin du Leman. *Revue suisse de Zoologie* **23**.
Morgan, C.I. & King, P.E. (1976). *British Tardigrades.* London: Academic Press.
Murray, J. (1905). On a new family and 12 new species of the order Bdelloidea, collected by the Lake Survey. *Transactions of the Royal Society of Edinburgh* **41**, 367–386.
Murray, J. (1908a). *Philodina macrostyla* Ehrenberg and its allies. *Journal of the Quekett Microscopical Club*, Series **II, 10,** 207–226.
Murray, J. (1908b). Scottish rotifers collected by the Lake Survey (supplement). *Transactions of the Royal Society of Edinburgh* **46,** 189–201.
Norgaard, E. (1956). On the ecology of two lycosid spiders (*Pirata piratica* and *Lycosa pullata*) from a Danish *Sphagnum* bog. *Oikos* **3**, 1–21.
Page, F.R. (1976). *An Illustrated Key to Freshwater and Soil Amoebae.* Freshwater Biological Association Scientific Publications **34**.
Pascher, A. (ed.) (1913–1927). *Die Süsswasser-Flora Deutschlands, Österreichs und der Schweiz.* Jena: G. Fischer.
Pascher, A. (1930a). Ein grüner *Sphagnum*-epiphyt und seine Bezeitung zu freilebenden Verwandten. *Archiv für Protistenkunde* **69**, 637–658.
Pascher, A. (1930b). Über einen grünen assimilationsfähigen plasmodialen organismus in den Blättern von *Sphagnum. Archiv für Protistenkunde* **72,** 311–358.

Pascher, A. (1932). Über eine in ihrer Jugend rhizopodial und animalisch lebende epiphytische Alge (*Perone*). *Beihefte zum Botanische Zentralblatt* **49**.

Pascher, A. (1937). *Kryptogrammenflora von Deutschland, Österreich und der Schweiz. Heterokonten*, XI, 256–275. Leipzig: Akademische Verlagsgesellschaft.

Pascher, A. (1940). Rhizopodiale Chrysophyceen. *Archiv für Protistenkunde* **93**, 331–349.

Pearsall, W.A. (1950). *Mountains and Moorlands*. London: Collins.

Pennak, R.W. (1978). *Freshwater Invertebrates of the United States* (2nd edn). New York: John Wiley & Sons.

Pennak, R.W. (1989). *Freshwater Invertebrates of the United States* (3rd edn). *1. Protozoa to Mollusca*. New York; John Wiley & Sons.

Pentecost, A. (1982). The distribution of *Euglena mutabilis* in sphagna with reference to the Malham Tarn North Fen. *Field Studies* **5**, 591–606.

*Pentecost, A. (1984). *Introduction to Freshwater Algae*. Slough: The Richmond Publishing Co.Ltd

Pontin, R.M. (1978). *A Key to British Freshwater Planktonic Rotifera*. Freshwater Biological Association Scientific Publications **38**.

Prescott, G.W. (1962). *Algae of the Great Western Lakes Area* (3rd edn). Michigan: Cranbrook Institute of Science.

Rainer, H. (1968). Heliozoa. *Die Tierwelt Deutschlands*, **56.**

Richardson, D.H.S. (1981). *The Biology of Mosses*. Oxford: Blackwell Scientific Publications.

Rixen, J.U. (1961). Kleineturbellarian aus dem litoral der Binnengewasser Schleswig-Holsteins. *Archiv für Hydrobiologie* **57**, 464–538.

Rogerson, A. (1982). An estimate of the annual production and energy flow of the large naked amoeba population inhabiting a *Sphagnum* bog. *Archiv für Protistenkunde* **126**, 146–149.

Rose, F. (1953). A survey of the ecology of British lowland bogs. *Proceedings of the Linnean Society* **164**, 186–212.

Round, F.E. (1981). *The Ecology of Algae*. Cambridge: Cambridge University Press.

Round, F.E., Crawford, R.M. & Mann, D. (1990). *The Diatoms*. Cambridge: Cambridge University Press.

Ruttner-Kolisko, A. (1974). Planktonic rotifers. Biology and taxonomy. *Die Binnengewässer* (supplement) **26**, 1–146.

Scourfield, D.J. & Harding, J.P. (1966). *A Key to the British Species of Cladocera*. Freshwater Biological Association Scientific Publications **5**.

Smith, A.J.E. (1978). *The Moss Flora of Britain and Ireland*. Cambridge: Cambridge University Press.

Smith, A.J.E. (ed.) (1982). *Bryophyte Ecology*. London: Chapman & Hall.

Springett, J.A. (1970). The distribution and life histories of some moorland Enchytraeidae (Oligochaeta). *Journal of Animal Ecology* **39**, 725–737.

Standen, V. & Latter, P.M. (1977). Distribution of a population of *Cognettia sphagnetorum* (Enchytraeidae) in relation to micro-habitats in a blanket bog. *Journal of Animal Ecology* **46**, 213–229.

Stewart, W.D.P. (1966). *Nitrogen Fixation in Plants*. London: Athlone Press.

Stewart, W.D.P (1973). Nitrogen fixation by photosynthetic micro-organisms. *Annual Review of Microbiology* **27**, 283–316.

Tarras-Wahlberg, N. (1952). Oribatids of the Akhult Mire. *Oikos* **6**, 166–171.

Tarras-Wahlberg, N. (1962). The Oribatei of a Central Swedish bog and their environment. *Oikos*, Supplement **4**, 1–56.

Taylor, F.J.R. (ed.) (1987). *The Biology of the Dinoflagellates*. Oxford: Blackwell Scientific Publications.

Thorp, J.H. & Covich, A.P. (1991). *Ecology and classification of North American Freshwater Invertebrates*. San Diego: Academic Press.

*Unwin, D.M. & Corbet, S.A. (1991). *Insects, Plants and Microclimate*. Naturalists' Handbooks No. **15**. Slough: The Richmond Publishing Co. Ltd.

Ward, H.B. & Whipple, G.C.(1918). *Freshwater Biology*. Boston: Stanhope Press.

West, G.S. & Fritsch, F.E. (1927). *A Treatise on the British Freshwater Algae*. Cambridge: Cambridge University Press.

Wright, J.C. (1989). Desiccation tolerance and water-retentive mechanisms in tardigrades. *Journal of Experimental Biology* **142**, 267–292.

Young, J.O. (1970). British & Irish freshwater microturbellaria: historical records, new records and a key for their identification. *Archiv für Hydrobiologie* **67**, 210–241.

Voigt, M. (1956). *Rotatoria. Die Rädertiere Mitteleuropas*. Berlin: Borntraeger.

Voigt, M. (1960). Gastrotricha. *Die Tierwelt Mitteleuropas* **1, 4a**, 1–74.

Index